Writing Science

Writing Science

How to Write Papers That Get Cited and
Proposals That Get Funded

JOSHUA SCHIMEL

OXFORD
UNIVERSITY PRESS

Oxford University Press, Inc., publishes works that further
Oxford University's objective of excellence
in research, scholarship, and education.

Oxford   New York
Auckland   Cape Town   Dar es Salaam   Hong Kong   Karachi
Kuala Lumpur   Madrid   Melbourne   Mexico City   Nairobi
New Delhi   Shanghai   Taipei   Toronto

With offices in
Argentina   Austria   Brazil   Chile   Czech Republic   France   Greece
Guatemala   Hungary   Italy   Japan   Poland   Portugal   Singapore
South Korea   Switzerland   Thailand   Turkey   Ukraine   Vietnam

Published by Oxford University Press, Inc.
198 Madison Avenue, New York, New York 10016
www.oup.com

Oxford is a registered trademark of Oxford University Press

Library of Congress Cataloging-in-Publication Data

Schimel, Joshua.
 Writing science : how to write papers that get cited and proposals that get funded / Joshua Schimel.
    p. cm.
 Includes bibliographical references and index.
  ISBN 978-0-19-976023-7 (hardcover : alk. paper) — ISBN 978-0-19-976024-4 (pbk. : alk. paper)
 1. Technical writing. 2. Proposal writing for grants. I. Title.

 T11.S35 2012
 808.06'65—dc23

                                    2011028465

18

Printed in the United States of America on acid-free paper

To my father, Jack Schimel, who loved language

CONTENTS

Preface ix
Acknowledgments xiii

1 Writing in Science 3

2 Science Writing as Storytelling 8

3 Making a Story Sticky 16

4 Story Structure 26

5 The Opening 35

6 The Funnel: Connecting O and C 50

7 The Challenge 58

8 Action 67

9 The Resolution 83

10 Internal Structure 95

11 Paragraphs 104

12 Sentences 112

13 Flow 124

14 Energizing Writing 133

15 Words 145

16 Condensing 158

17 Putting it All Together: Real Editing 174

18 Dealing with Limitations 180

19 Writing Global Science 189

20 Writing for the Public 195

21 Resolution 204

Appendix A: My Answers to Revision Exercises 207
Appendix B: Writing Resources 212

Index 215

Those who can do, also teach.

It came as a surprise to me one day to discover that I was writing a book on writing. It's not the normal pastime for a working scientist, which I am—I'm a professor of soil microbiology and ecosystem ecology. I write proposals, I write papers, and I train students to do both. I review extensively and have served as editor for several leading journals. Teaching writing evolved from those activities, and it became a hobby and a passion. This book is the outgrowth—it's what I have been doing when I should have been writing papers.

Although I believe I have become a good writer, I got there through hard work and hard lessons. I didn't start out my academic life that way. Before teaching my graduate class on writing science for the first time, I went back to my doctoral dissertation for a calibration check—what should I expect from students? I made it through page 2. At that point, my tolerance for my own writing hit bottom and my appreciation for my advisor's patience hit top. Even the papers those clumsy chapters morphed into were only competent.

My writing has improved because I worked on becoming a writer. That doesn't mean just writing a lot. You can do something for many years without becoming competent. Case in point: the contractor who put a sunroom on our house. He kept insisting, "I've been doing this for 20 years and know what I'm doing"; the building inspector's report, however, said to reframe according to building codes and standard building practices.

I have learned to write through a number of avenues: guidance from my mentors; the trial and error of reviews and rejections; thinking about communication strategy; working with students on their papers; reviewing and editing hundreds of manuscripts; reading and rereading books on writing; and importantly, participating in my wife's experiences as a developing writer, listening to the lessons from her classes, and watching how real writers train and develop. I have tried to meld all these lessons into science writing, incorporating writers' perspectives into the traditions and formulas of science. This book represents that amalgamation, and I hope it will help you short-circuit the long, slow, struggle I experienced.

PRINCIPLES VERSUS RULES

Many books on writing (notably the bad ones) present a long string of rules for how to write well. In them, writing is formulaic. In good writing, however, "the code is more what you call guidelines than actual rules" (to quote from *Pirates of the Caribbean*), a point made strongly by two prominent writers on writing, Joseph Williams (*Style: Toward Clarity and Grace*) and Roy Peter Clark (*The Glamour of Grammar*). Most of the time, following the rules will improve your writing, but good writers break them when it serves their purposes. I distinguish such rules from principles, which are the general concepts that guide successful communication. If you violate principles, your writing *will* suffer.

Throughout the book I try to distinguish between rules and principles, and I hope to offer enough insight that you will understand which are which, and why. When following a rule conflicts with following a principle, flout the rule freely and joyously.

SOURCES FOR EXAMPLES

I found examples in many places—some from work I know, some from papers that friends recommended, one from someone I met on an airplane, and many from randomly flipping through journals. The examples I hold up as good practice, I use intact and cite properly, though I remove the reference citations to make them easier to read. Exemplars of good practice deserve to be recognized. I sometimes point out what I see as imperfections, but only to highlight that even good writing can usually be better, and that although we may strive for perfection, we never reach it. A "good enough" proposal may still get funded, and an award letter from the National Science Foundation is the best review I've ever seen.

The examples of what I think you should *not* do are closely modeled on real examples. However, unless they come from my own work, I have rewritten the text to mask the source. When I rewrote the text, I maintained the structural problems so that even if the science is no longer "real," the writing is. In some cases these examples are from published work; in others, from early drafts that were revised and polished before publication. If you recognize your own writing or my comments on it (if I had handled it as a reviewer or editor), please accept my thanks for stimulating ideas that I could use to help future writers. We learn from our mistakes, and I need to show readers real "mistakes" to learn from. I hope I helped with the reviews I wrote at the time.

When I take examples from my own work, it is because only then can I accurately explain the author's thinking. When I use others' work, I can assess what they did and why it worked or failed, but I can't know why they made the choices they did. For proposals, I use my own extensively because I have access to them. Proposals aren't published, so I can't scan other fields to find good examples, as I could for papers.

I have included examples from many scientific disciplines to illustrate that my approaches and perspectives are broad-based; the basic challenges and strategies of writing are similar across fields. Many, however, come from the environmental sciences, where I knew where to find useful examples and where I felt that most readers would be able to understand enough of the content to have an easier time focusing on the writing.

EXERCISES AND PRACTICE

In most chapters, I include exercises to apply the concepts I discuss. I encourage you to work through these, ideally in small groups. Writers often have writer's groups, where typically four to six people get together to work over each other's material, discuss what works and what doesn't, and suggest alternative ways of doing things. This process is helpful in developing successful writers—it provides insights from different points of view that can stretch boundaries and offer new ideas. Analyzing others' work can hone analytical skills. Groups also provide a supportive environment for learning, analogous to how a lab group helps you expand your research tools.

The exercises fall into several categories. The most important is the short article I ask you to write (and rewrite, and then rewrite again). I use this exercise in my writing class, and it is enormously successful, particularly when coupled with peer discussion and editing. The short form intensifies the focus on the story as well on each paragraph and sentence.

The second important exercise is to analyze the writing in published papers. How did the authors tell their story? Did it work? Was it clear? How could you improve the writing? This, too, is best done in groups. These papers don't need to be the best writing in the field—we can learn as much from imperfect writing as we do from excellent work. The rule in these discussions should be that you may not discuss the scientific content unless it is directly germane to evaluating the writing. Get in the habit of evaluating the writing in every paper you read or discuss— the more you sensitize yourself, the more those insights will diffuse into your own writing.

Finally, there are editing exercises that target specific issues such as sentence structure, word use, and language. For those, I provide suggested answers at the back of the book. Remember, though, that there is never a single way to approach a writing problem; my answers are not the only approach and may not even be the best. In working examples in class, students often find different and better solutions than any I came up with.

If you really want to become a better writer, do the exercises. Work with your friends and colleagues on them. You only learn to write by writing, being edited, and rewriting. You must learn not just the principles but also how to apply them.

The point is that you have to strip your writing down before you can build it back up. You must know what the essential tools are and what job they were designed to do. Extending the metaphor of carpentry, it's first necessary to be able to saw wood neatly and to drive nails. Later you can bevel the edges or add elegant finials, if that's your taste. But you can never forget that you are practicing a craft that's based on certain principles. If the nails are weak, your house will collapse. If your verbs are weak and your syntax is rickety, your sentences will fall apart.

WILLIAM ZINSSER, *On Writing Well*

ACKNOWLEDGMENTS

I always blame this book on Christina Kaiser and Hildegard Meyer, two graduate students at the University of Vienna. But the person really responsible, as she is for most of the best things in my life, is my wife, Gwen. We spent the summer of 2005 in Montpellier, France, at the Centre d'Ecologie Fonctionnelle et Evolutive of the CNRS, hosted by Stefan Hättenschwiler and Giles Pinay; we took the opportunity to go to Vienna to visit Dr. Andreas Richter and his research group. Tina and Hildegard were chatting with Gwen and mentioned that they liked reading my papers because they were well written. That sparked Gwen to suggest I teach a workshop on writing for the lab group in France. The rest is history. So Tina and Hildegard, little may you realize the power of that off-hand comment, but you catalyzed this. Thank you.

My thanks to Gwen are endless—not only did teaching writing come from her inspiration, but much of what I know about writing and how writers learn their craft comes from her. She supported and encouraged me through the years I've worked on this, and she has read through most of the book, providing valuable insights and feedback.

The other critical thread that led to my writing this book was becoming a 2006 Aldo Leopold Leadership Fellow. Not only was the Leopold program's communication training influential, but simply being a fellow helped motivate me to take what I had learned and make it available to the community.

Many of my colleagues have given me ideas, insights, quotes, and good stories about science and communication. Many of those comments were made in passing and were not targeted at either writing or this book. You may not realize how sticky those ideas were, and you may not even remember saying them, but thank you. I have been privileged to work with as talented, insightful, and generous a group of friends and colleagues as I can imagine. I am grateful to you all for enriching my work and my life.

Many of those colleagues have reviewed my work over the years and forced me to develop my writing and thinking skills to get proposals funded and papers published. At the time, I may have complained about that "miserable know-nothing so-and-so," and I once commented about a good friend who was the editor handling a paper that "If he accepts this version, I owe him a beer; if he sends it back for more revision, I'm going to pour it on him." I am, however, grateful to you all for holding my feet to the fire and forcing me to make my work as good as it could be. It both built my scientific career and taught me how to write.

My Ph.D. advisor, Mary Firestone, taught me the most crucial lessons of how to frame the question and the story. When I was finishing my dissertation, she also edited my horrible, sleep-deprived writing into a form that was at least minimally acceptable and did so with grace and humor. She set me on this path.

Erika Engelhaupt gave me great suggestions and great text for chapter 20, "Writing for the Public." Weixin Cheng provided valuable suggestions on chapter 19, "Writing Global Science." Bruce Mahall and Carla D'Antonio, with whom I lead the Tuesday evening plant and ecosystem ecology seminar, have helped me deepen my insights into communication strategy. Carin Coulon drew the wonderful figure of the Roman god Janus that appears in chapter 13.

I owe great thanks to the U.S. National Science Foundation. The NSF is an extraordinary organization, due to the talent and dedication of its program officers. The NSF has supported my work and helped me grow to reach the point where I could write this book.

Many people have participated in the workshops I've given on writing and in the graduate class I teach. This book grew from them, and in working through the lessons in person I have been able to polish them. Thank you all.

I've worked on manuscripts with a number of graduate students and postdocs. They helped me develop my own writing tools and my analytical understanding of those tools so I could teach them to others. The list is long and grows longer monthly: Jay Gulledge, Mitch Wagener, Joy Clein, Jeff Chambers, Mike Weintraub, Noah Fierer, Sophie Parker, Doug Dornelles, Shawna McMahan, Shinichi Asao, Izaya Numata, Ben Colman, Knut Kielland, Susan Sugai, Carl Mikan, Andy Allen, Michael LaMontagne, Amy Miller, Matt Wallenstein, Shurong Xiang, Dad Roux-Michollet, Sean Schaeffer, Claudia Boot, Mariah Carbone, and Yuan Ge. Particular thanks go to Shelly Cole for her generosity. Thanks also to all the other students whose dissertations and manuscripts I have read and edited while serving on your committees.

Finally, I would like to note two books that have greatly influenced my thinking on writing and communication: Joseph Williams's *Style: Toward Clarity and Grace*, and Chip and Dan Heath's *Made to Stick*. Williams's book is the best book on writing I have ever read, and I am deeply indebted to him for his insights, many of which I have assimilated into this book (filtered through my own experiences and focused on writing science). I cannot match his insights into the sophistication of the English language, so I recommend that you reread it regularly and give copies to your friends and students. *Made to Stick* isn't ostensibly about writing at all, and distinctly it isn't about writing science. Rather, it focuses on advertising, marketing, and general communication. It is, however, a spectacularly insightful and fun discussion of what makes ideas engaging and "sticky," a critical issue for scientists who want their work to get noticed from among the overwhelming flood of papers published every year.

Writing Science

Writing in Science

As a scientist, you are a professional writer.

Success as a scientist is not simply a function of the quality of the ideas we hold in our heads, or of the data we hold in our hands, but also of the language we use to describe them. We all understand that "publish or perish" is real and dominates our professional lives. But "publish or perish" is about surviving, not succeeding. You don't succeed as a scientist by getting papers *published*. You succeed as a scientist by getting them *cited*.

Having your work matter, matters. Success is defined not by the number of pages you have in print but by their influence. You succeed when your peers understand your work and use it to motivate their own. The importance of citation and impact is why journals measure themselves by the Impact Factor and why the citation-based H-factor is becoming more important for evaluating individual researchers. If you have 10 publications that have each been cited 10 times, you have an H of 10; if you have 30 papers that have each been cited 30 times, you have an H of 30; but if you have published 100 papers and none have been cited, on the H-factor you would rate a flat zero. Success, therefore, comes not from writing but from writing effectively.

The power of writing well also explains a pattern I noticed as I was looking for examples to include in this book, a pattern I had only been unconsciously aware of before. When I needed examples of good writing, I could usually go to the leaders in various fields—most write exceptionally well. They are able to cast their ideas in language that is clear and effective and that communicates to a wide audience. Is this pattern accidental? I doubt it. These men and women not only think more deeply and creatively than most of us, they also are able to communicate their thinking in ways that make it easy to assimilate. That is how they became leaders.

Your initial reaction to this observation may be to assume that these people think more clearly than most, and thus they write more clearly. Certainly they do both, but it is less obvious which way causality goes. Does clear thinking lead to clear writing? Or, alternatively, does clear writing lead to clear thinking? The answers to these questions may seem intuitive, but they are not.

> I ask, finally, that you avoid one error of belief that is monstrously prevalent. This is the widespread notion that "to write clearly, you must first think clearly." This sharp little maxim may appear logical, but it is really rubbish. No matter how rational your thought may be (or appear to be) on a particular problem, no matter how detailed your intentions and plottings, the act of writing will almost always prove rebellious, full of unforeseen difficulties, sidetracks, blind alleys, revelations. Good, clear writing—writing that teaches and informs without confusion—emerges from a process of struggle, or if you prefer, litigation.
>
> Most often, the terms of the formula given above need to be reversed: "clear thinking can emerge from clear writing." Imposing order by organizing and expressing ideas has great power to clarify. In many cases, writing is the process through which scientists come to understand the real form and implications of their work.
>
> SCOTT MONTGOMERY. *The Chicago Guide to Communicating Science*[1]

I agree with Montgomery. Often, the process of structuring your thoughts to communicate them allows you to test and refine those thoughts. As you focus on writing clearly, you force yourself to think more clearly. Improving your writing will help you become successful, both because it allows you to communicate your ideas more effectively, making them accessible to the widest audience, and also because it makes your thinking, and thus your science, better.

This brings me back to my original argument—as a scientist, you are a professional writer. Writing is as important a tool in your toolbox as molecular biology, chemical analysis, statistics, or other purely "scientific" tools. Some of these tools allow us to generate data; others to analyze and communicate results. Writing is the most important of the latter. Because it forms the bridge to your audience,

---

1. S. L. Montgomery, *The Chicago Guide to Communicating Science* (University of Chicago Press, 2003).

it can act as the rate-limiting step that constrains the effectiveness of all the other tools.

Despite the importance of writing, however, for most scientists it is something we do post hoc. After we get the data, we "write up" the paper. This is an unfortunate approach. Because writing is a critical tool, you should study it and develop it as thoroughly as your other tools. Writing is as complex and subtle as molecular biology.

> I wish I had a secret I could let you in on, some formula my father passed on to me in a whisper just before he died, some code word that has enabled me to sit at my desk and land flights of creative inspiration like an air-traffic controller. But I don't. All I know is that the process is pretty much the same for almost everyone I know. The good news is that some days it feels like you just have to keep getting out of your own way so that whatever it is that wants to be written can use you to write it.
>
> But the bad news is that if you're at all like me, you'll probably read over what you've written and spend the rest of the day obsessing, and praying that you do not die before you can completely rewrite or destroy what you have written, lest the eagerly waiting world learn how bad your first drafts are.
>
> ANNE LAMOTT, *Bird by Bird*[2]

Even the most successful writers struggle with writing. It is actually easier for us as scientist writers because as readers, our expectations are low and we want the information—we'll fight through cluttered sentences and disconnected paragraphs to try to get it. But if readers have to fight that fight, some will lose, and then you, the author, will be the greater loser. How many papers are so brilliant, so earth-shattering, so discipline-changing that if you don't read and assimilate them, your research will be blighted and your career will suffer? Do you need more than the fingers on one hand to count them? Most of us never write one. Rather, we build our careers incrementally—our peers read our papers and use our ideas; the more papers we publish and the more they are used, the more successful we are. But our work gets read and cited because we made our points well enough that readers could follow what we were saying. Our proposals are funded because we were able to make our ideas clear, compelling, and convincing to reviewers. Our success, then, comes from our ability to communicate our ideas as much as from their inherent quality. As the author, therefore, your job is to make the reader's job easy.

That last point may be the overriding principle that all the others in this book grow out of, so let me repeat it, louder. *It is the author's job to make the reader's job easy.*

Despite the importance of writing effectively, many respected scientists are at best only competent writers, and we could all be better. Yet most books on science writing take a technical approach to preparing a manuscript, focusing on basic

2. A. Lamott, *Bird by Bird* (Anchor Books, 1994).

information such as how to structure a paper, whether to use figures or tables, and how to manage the process of submitting a paper and dealing with editors and reviewers. Those books are more about publishing than about writing; they treat writing as something a scientist *does*.

I take a different approach—treating *being a writer* as something a scientist *is*. That distinction may appear subtle, but it is profound. If writing is merely something you do, like washing the glassware after an experiment—a perhaps unpleasant afterthought—you will never be a successful writer. You will not invest in sharpening your tools or expanding your toolbox; you may not be aware that you even have a "writing toolbox." That changes when you recognize that you are a writer and accept it as your profession. Professionals pay attention to their craft, study it, analyze the work of peers to learn from them, develop new tools, and experiment with new approaches. They grow in their ability to perform with style and power, whether that be to create wooden chairs, legal arguments, life-saving surgeries, or scientific papers that become classics. If you want your writing to be effective, become a writer.

This book is unapologetically on the craft of writing—communicating through the written word. I won't tell you how to put together a figure, how to assemble a bibliography, or how to decide where to submit the paper. There are excellent books that cover that material, and I intend this book to complement rather than replace them. Instead, I target scientists—from students to working professionals—who are ready to go beyond the basics and become writers.

While focusing on the specific issues we face as scientists in producing papers and proposals, I approach the challenge of technical writing from the perspective of a writer, thinking about the issues the way professional writers do. Thus, a large part of the book is about story and story structure—how you lay out issues, arguments, and conclusions in a coherent way. If you can't deal with the big issues, the small ones don't matter very much. Good tactics never overcome bad strategy. Then I move on to finer scales, from overall story structure through paragraphs and sentences to how we choose individual words. The final section covers specific challenges that arise in different types of science writing.

1.1. WRITING VERSUS REWRITING

One thing to keep in mind as you read this book and apply the ideas to your own work is that this is really a book about rewriting, not writing. Every single thing I tell you not to do, I do in my first drafts—I may do them less than I used to, but I still do them. First drafts, though, don't matter; no one else sees them. Trying to get a first draft perfect can be paralyzing, a phenomenon well recognized by the best writers on writing.

A warning: if you think about these principles *as you draft*, you may never draft anything. Most experienced writers get something down on paper or up on the screen as fast as they can, just to have something to revise. Then as they rewrite an earlier draft into something clearer, they more clearly understand their ideas. And when they understand their ideas better; they express them more clearly, and when they express them more clearly, they understand them even better . . . and so it goes, until they run out of energy, interest, or time.

JOSEPH WILLIAMS, Style: Ten lessons in clarity and grace[3]

Rewriting is the essence of writing. I pointed out the professional writers rewrite their sentences over and over and then rewrite what they have rewritten.

WILLIAM ZINSSER, On Writing Well[4]

The last word on rewriting comes from Anne Lamott, who addresses it with humor and insight:

> Shitty First Drafts. All good writers write them. That is how they end up with good second drafts and terrific third drafts.
>
> I know some very great writers, writers you love who write beautifully and have made a great deal of money, and not one of them sits down routinely feeling wildly enthusiastic and confident. Not *one* of them writes elegant first drafts. All right, one of them does, but we do not like her very much.

Unfortunately, this quote highlights just how wonderful a writer Lamott is—her third drafts are terrific. When I finish a paper, there are usually 10 or 20 drafts cluttering up my computer, and I only think the last one is terrific until I reread it later. Rereading things I've written is often painful; imperfections glow like neon signs, leaving me to wonder how I ever managed to miss them in the first place.

Writing can be a painful process of rewriting, rewriting, and more rewriting until your work gets good enough to send off. An artist never completes a work—they merely let it go. This rewriting cycle develops both your writing and your thinking, moving both toward clarity and power. How do you get to Carnegie Hall? Practice, practice, practice! How do you get an award letter from the National Science Foundation or the National Institutes of Health? Polish, polish, polish! If you are going to be a successful writer, learn to embrace the pain and enjoy the process.

3. J. M. Williams, *Ten Lessons in Clarity and Grace* (Longman, 2005).

4. W. Zinsser, *On Writing Well* (HarperCollins, 1976).

2

Science Writing as Storytelling

A good story cannot be devised; it has to be distilled.

—RAYMOND CHANDLER

Elizabeth Kolbert, the author of the extraordinary book on climate change *Field Notes from a Catastrophe*, once said that the problem she has with scientists is that we don't tell stories. That statement bothered me, because we do. If we didn't tell stories, we would write papers with only Methods and Results; we could skip the Introduction and Discussion. We also wouldn't read Charles Darwin's *Origin of Species*; instead, we would read his notebooks and get the raw data.

But, we do write papers with an Introduction and a Discussion, and we do read *Origin of Species*. A paper doesn't only present our data—it also interprets them. A paper tells a story about nature and how it works; it builds the story from the data but the data are not the story. The papers that get cited the most and the proposals that get funded are those that tell the most compelling stories.

Somehow, though, our kind of storytelling didn't connect with Kolbert; in fact, it connected so poorly that she didn't recognize our stories as stories. Why? I suspect three reasons for this. First, scientists tell stories using a formalized structure that doesn't match well with that used by journalists. Our stories get lost in the struggle of cross-cultural communication. Second, many of us are poor storytellers; either we don't see the story clearly or we just can't tell it clearly. Finally, some (perhaps most) scientists are uncomfortable with thinking about what we do as "telling stories." Many associate the idea of "stories" with fiction. Scientists are supposed to be objective and dispassionate. Arguing that you are writing a story may seem to suggest that you have left that objectivity behind and with it, your professionalism. Rather, many scientists feel that their job is simply to "present their work," and so do a poor job of highlighting the story. The result is that even an outstanding journalist who spends a lot of time talking with scientists doesn't recognize that we *are* telling stories.

That lack of recognition raises several issues that scientists should consider. The first is the formalism of how we write papers and proposals. I won't argue that we should change how we structure these documents; they serve our needs to communicate among ourselves. (The phenomenon that they don't communicate well to the rest of the world is a different concern.) The second issue is how to become better storytellers and better communicators. That is something we can all work on.

The final issue is more complex. Is seeing science writing as storytelling professional or not? Journalists are also supposed to be objective and dispassionate (and the best ones are), yet their entire discipline is grounded in the concept of "story." So there is nothing inherently unobjective or unprofessional in the idea of storytelling. To tell a good story in science, you must assess your data and evaluate the possible explanations—which are most consistent with existing knowledge and theory? The story grows organically from the data and is objective, dispassionate, and fully professional. Where you run into problems is when the authors know the story they want to tell before they collect the data and then try to jam those data into that framework. Anne Lamott captures this conundrum well. Although she was discussing fiction, her advice applies equally to science.

> Characters should not, conversely, serve as pawns for some plot you've dreamed up. Any plot you impose on your characters will be onomatopoetic: PLOT. I say don't worry about plot. Worry about the characters. Let what they say or do reveal who they are, and be involved in their lives, and keep asking yourself, Now what happens?"
>
> ANNE LAMOTT, *Bird by Bird*

Lamott highlights the importance of listening to your characters to draw the story out of them, rather than imposing it on them. How do we, as scientists, take this advice? Do we even have "characters" to listen to? Of course we do. Our characters, however, aren't people; instead, they may be molecules, organisms, ecosystems, or concepts. Nitrogen cycling in the arctic tundra, benzene and its reactions, or genes and their functions can be characters that we "listen to" by carefully analyzing our data with an open mind. Then we can develop these characters in a paper as we discuss them and what makes them tick.

Kolbert's difficulty with understanding our stories raises the social imperative of our becoming better storytellers. As science has moved from esoteric, ivory-tower natural philosophy to something that directly affects the lives and well-being of the public, our inability to communicate has grown into a crisis. Science is often ignored, misunderstood, or misrepresented in the public arena and in policy decisions, a phenomenon many of us bemoan. How can we solve problems as serious as global warming or cancer without basing the solutions on the best available science? Ensuring that science is used properly requires more than just presenting facts to decision makers. Unfortunately, our approach to communicating to them is often analogous to traveling overseas and speaking louder when the locals don't understand English. Going to Washington, D.C. and speaking loudly to the locals in "science" is about as successful—it doesn't get our point across, and it makes us seem arrogant, a good way to get dismissed. Our inability to communicate outside the narrow confines of our specializations undermines our ability to influence policy and to generate new sources of funding. We don't have to become science popularizers like Stephen Jay Gould or Carl Sagan, we just have to become better storytellers. Doing so will make us more effective with each other, with our professional translators (science journalists like Kolbert), with policy makers, and with the public.

2.1. FINDING THE STORY

The distinction between presenting results and telling a story embodies a challenge for many when writing papers. If you believe that writing a paper is about presenting results, then it would seem reasonable to outline everything you did and then say something about it. But somewhere in that mass of data is a story trying to come out. Find it, and give it to us.

In looking for the story, remember that when we do science, we get data from the mass spectrometer, the DNA sequencer, or the telescope, but our ultimate goal is not those data –it is the understanding we derive from them. In the discovery of the structure of DNA and the molecular basis for heredity, it wasn't Rosalind Franklin's Photo 51,[1] the critical X-ray diffraction image of DNA (figure 2.1a) that gained fame but the sketch of the molecular structure of DNA that Francis Watson and James Crick built from it (figure 2.1b).[2] Franklin's lack of credit for her role in the discovery has created controversy over the years because there can be no story without the underlying data, but that controversy is a separate issue. My point is that raw data have limited direct value and are usually interpretable by only a small group of experts—Photo 51 means nothing to me beyond its role as a historical artifact. The double helix model of DNA, however, I understand. It is interpretable by many and is at the core of the work of thousands of scientists spanning from medicine to soil microbiology. Watson and Crick's groundbreaking paper

1. R. E. Franklin and R. G. Gosling, "Molecular Configuration in Sodium Thymonucleate," *Nature* 171 (1953): 740–41.

2. J. D. Watson and F. H. C. Crick, "Molecular Structure of Nucleic Acids—A Structure for Deoxyribose Nucleic Acid," *Nature* 171 (1953): 737–38.

A. Photo 51

B. Model of
DNA

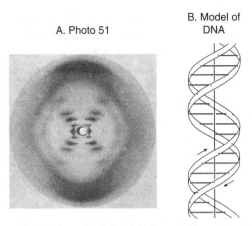

Figure 2.1. Photo 51, Rosalind Franklin's critical X-ray diffraction image of crystallized DNA and the simple model of its structure developed by James Watson and Francis Crick.
Both images © 1953, Nature Publishing Group, reprinted with permission.

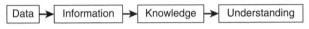

Figure 2.2. The flow of science, from data to understanding.

had power because they used the data to tell a story about nature and how it works, developing an intellectual model of DNA structure and what that implies for heredity. We look for and value such insights and understanding.

The role of scientists is to collect data and transform them into understanding. Their role as authors is to present that understanding. However, going from data to understanding is a multi-step process (figure 2.2). The raw data that come from an instrument need to be converted into information, which is then transformed into knowledge, which in turn is synthesized and used to produce understanding. In the case of DNA, Photo 51 was data—an image of X-ray scattering. Franklin used that data to produce actual, critical information on the atomic structure of crystallized DNA. Watson and Crick used that information to produce knowledge—the double helix structure. The last step is understanding—taking that knowledge about the molecule's structure to explain how it allows cell replication and heredity.

The further along the path from data to understanding you can take your work and your papers, the more people will be able to assimilate your contributions and use them to motivate their own work and ideas—and that should be your goal. If you don't provide understanding (or at least knowledge) readers will be left searching for it. The data are supporting actors in the story you tell. The lead actors are the questions and the larger issues you are addressing. The story grows from the data, but the data are not the story.

This recognition leads to a process that I think is critical to developing good stories and writing good papers, a process that hearkens back to Lamott's comments about listening to your characters: develop your story from the bottom up, then tell it from the top down. Start with the data, think about them, listen for the story they are trying to tell, and find that story. Don't listen just to your characters' loud proclamations, though; listen also to their quiet, uncertain mutterings. What might that shoulder on the spectrum mean? If that nonsignificant treatment effect were real, what would that say about your system? Is that outlier a flag for something you hadn't thought about but may be important? Overinterpret your data wildly, and consider what they might mean at those farthest fringes. Explore the possibilities and develop the story expansively. Then, take Occam's razor and slash away to find the simple core.

Why go through this "elaborate and slash" process? Isn't elaborating a waste of time if you're going to come back to a simple story in the end? Why not start there? Well, if you start with the first simple story that comes to mind, you are probably imposing plot onto your characters and falling into the trap Lamott describes. Only by exploring the boundaries and limits of your data can you find the important story.

The power of the exploring the fringes is well illustrated by Bill Dietrich's graduate research. Dietrich is now a professor of geomorphology at the University of California, Berkeley, and is a member of the U.S. National Academy of Sciences. For his doctorate, he worked on how hill slope steepness controlled soil depth in the Pacific Northwest. Most of the data fit a nice tight relationship (figure 2.3), which made a perfectly good story.[3] But there were outliers where soils were much deeper than they "should" be. He could have ignored them and focused on the main story. He didn't. He looked at the deep soils and what created them; he found that along a hill slope, the bedrock is uneven and in places forms hollow "wedges" (figure 2.3). Over time, those wedges fill up with debris and soil. Once filled, they aren't obvious on the landscape, but woe to the person who buys a house below one—in a heavy rainstorm, they can flush out, creating lethal mud flows. Evaluating the processes that fill and flush these wedges became a focus of Dietrich's early research career. Because he listened to his characters carefully, recognized that the most important story wasn't in the average but in the outliers, and then explored those outliers, he came up with more novel, exciting, and important science.

Learning to explore the fringes of your data, however, can be difficult and frustrating. When I was a graduate student, I would sometimes go to my advisor, Mary Firestone, with what I thought was a simple question. Then we might spend weeks discussing issues that wandered all over the intellectual map and didn't appear to fit on the straight road from my question to the answer. Many of the issues Mary raised seemed irrelevant and extraneous. What on Earth did the kinetics of bacterial glutamine synthetase have to do with my data on how plant roots compete against microorganisms for available nitrate in the soil? Over the years I worked with her, I came to understand what we were really doing in those conversations. Mary saw more of the system and how it fit together than

3. W. E. Dietrich and T. Dunne, "Sediment Budget for a Small Catchment in Mountainous Terrain," *Zeitschrift für Geomorphology* Suppl. Bd. 29 (1978): 191–206.

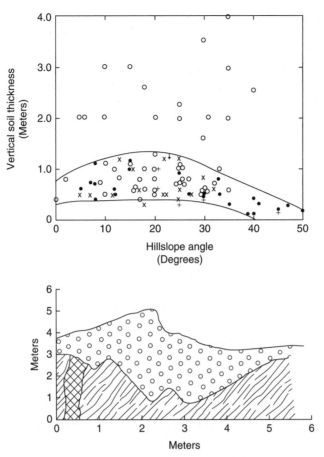

Figure 2.3. The top figure illustrates the relationship between hill steepness and soil depth in the U.S. Pacific Northwest; the bottom figure illustrates a cross-section through a wedge. Redrawn from Dietrich and Dunne (1978).
Copyright © 1978, E. Schweizerbart Science Publishers. www.borntraeger-cramer.de.
Reprinted with permission.

I did; she was teaching me how to do good science. She was exploring the issues to deepen our thinking, to ensure we found the story that tied together the sometimes apparently contradictory data, and to identify issues that might trip us up later. Though not always easy, it was an important lesson, one I remain grateful for.

So listen to your characters carefully—take the time to hear what they have to say and figure out what they mean. Fight the pressure to publish prematurely. One good paper can launch a career; many mediocre ones build a rather different one. Think well, write well, and then think some more while you write. Let the story grow from the data and then structure the paper to tell that story.

When we recognize that writing a paper is writing a story, it raises the obvious point that we can become better storytellers, better writers, and better scientists by studying what makes a good story, how other writers do it, and how to apply those ideas to science. We *can* communicate more effectively while remaining rigorously professional.

There are three aspects to effective storytelling. The first is content—what makes a story engage and stay with us? The second is structure—how do you put together that content to make it easy for us to get? The third is language—how do you write the story in the most compelling way possible? This book is about these three issues.

EXERCISES

2.1. Analyze published papers

Pick several papers from the primary literature. You will come back to these, chapter after chapter. I suggest you pick:

- A paper from a specialist journal written by a leader recognized as a strong writer.
- A "normal" paper from a specialist journal.
- A review or synthesis paper.
- A paper from *Nature* or *Science* or some journal that targets a broad audience.
- Identify what you think the key story points are. Did the authors do good job of highlighting that story? How far along the flow from data to understanding did the paper go? Could they have taken it further?

2.2. Write a short article

STEP 1. IDENTIFY THE KEY STORY POINTS FOR YOUR WORK.
(This is adapted from an exercise developed Ruth Yanai at SUNY-ESF.[4]) For each question, write a short paragraph—no more than two to three sentences. These identify the essential story elements.

1. What is your opening? This should identify the larger problem to which you are contributing, give readers a sense of the direction your paper is going, and make it clear why it is important. It should engage the widest audience practical.
2. What is your specific question or hypothesis?
3. What are the key results of your work? Identify these in a short list. There should be no more than two to three points.

---

4. She credits it to Bill Graves, Dick Gladon, and Mike Kelly at Iowa State University.

4. What is your main conclusion? What did you learn about nature? This should use the results from section 3 to answer the question from 2, and should address the larger problem identified in 1.

Step 2. Write the Article.

Write a short article describing your research. Your target audience is scientists who are not specialists in your discipline. You are trying to tell the story of your work and engage and educate your readers, not write a technical paper. The tone can range between somewhat technical and more casual, but it must be something that technical readers would find interesting. Use your answers from step 1 to frame the story you write in this part of the exercise.

The word limit is strict: 800–850 words.

Step 3. Analyze your writing.

Circulate your articles among your writers' group (a group of three to four people seems ideal for this). Analyze and edit each other's work. Then discuss the articles. Ask and answer the following questions:

1. What did the author do well? (It's always good to start positive.)
2. Was the topic interesting? Was it cast at the right level and hit the right audience? Could you have rewritten it to engage a wider audience? Did it make you want to read the rest of the piece?
3. Was the specific question clear?
4. Were the results clear? Did they relate to the topic and the specific question?
5. Were the Conclusions true *conclusions*, or were they merely a restatement of the results? Did they relate to the large issues raised in the opening? Did they answer the specific question asked? Did they clearly grow from the results presented in the piece?
6. What did you get as the "take-home" message of the story? Do you believe that this was the message the author was trying to give you?
7. Was the writing clear? If not, can you figure out why and identify ways to make it clearer?

3

Making a Story Sticky

A sticky idea is an idea that is more likely to make a difference.
—CHIP HEATH AND DAN HEATH

There are many ways to evaluate whether a story works, but perhaps the best is to ask, "How long after you read it do you remember it?" Some stories are riveting while you read but are gone as soon as you close the book. Perfect airplane reading. Others may stay with you for your entire life and be passed on to your children. Some are so powerful that they have lasted intact from the dawn of civilization.

Although nothing in science competes with the *Iliad* or the *Odyssey*, Darwin is still up there with his contemporaries Dickens and Dumas. Really good papers may be read and cited for years and decades. One of the nicest compliments I ever heard was someone saying a colleague wrote papers with "legs"—they stood the test of time, remaining interesting and relevant.

How do we write papers with legs—papers with immediate impact but that still accrue citations for years? In their book, *Made to Stick*,[1] Chip and Dan Heath frame this question as "What makes an idea 'sticky?" Why do some ideas stay

1. C. Heath and D. Heath, *Made to Stick* (Random House, 2007).

with you while others are eminently forgettable? Heath and Heath identify six factors that make an idea sticky and organize them in a simple mnemonic: SUCCES.

S: Simple
U: Unexpected
C: Concrete
C: Credible
E: Emotional
S: Stories

I go over these factors briefly here and come back to them repeatedly through the book. They are fundamental to good storytelling and thus to good science writing.

3.1. SIMPLE

Ideas that stick tend to be *simple*. A simple idea contains the core essence of an important idea in a clear compact way. Simple ideas have power.

During the U.S. Civil War, one of Abraham Lincoln's greatest challenges was dealing with antiwar Democrats, and in 1863 he faced a crisis. A leader of this faction, Clement Vallandigham, was preaching against the draft and encouraging soldiers to desert, undermining the war effort. He was arrested for treason, tried, and sentenced to prison. The fallout was furious. Was Lincoln using executive power to shut down the political opposition? Was Vallandigham just exercising his freedom of speech? The arguments were complex and impassioned. Lincoln cut through them all with a single question: "Must I shoot a simple-minded soldier boy who deserts, while I must not touch the hair of a wily agitator who induces him to desert?"

That question collapsed the complex legal and political arguments into a simple moral dilemma that people could understand and sympathize with. It made the innocent victim not Vallandigham but the soldier who listened to him and might pay the ultimate price for doing so. By framing the controversy in a simple, clear way, Lincoln refocused it and then shut it down. Bill Clinton was elected president on an even simpler message: "It's the economy, stupid."

It is important, however, to distinguish simple messages that capture the essence of an issue from those that are just "simplistic." Simplistic messages are dumbed down, trivialize the issue, or dodge the core of the problem, rather than targeting it. Many political slogans are simplistic; for example, "you pay too much in taxes" is catchy, appealing, and might even be true, but it ignores the underlying issues of what services those taxes pay for, whether you want or need them, and whether they provide good value for your money. Rather than condensing complex arguments about the balance of costs versus services, it avoids them— hence not simple, but simplistic.

Most science is driven by simple ideas. Frequently, the simpler an idea is at its core, the larger its swath of influence. Biology, for example, is driven by Darwin's theory of evolution by natural selection. Natural selection—fit organisms survive and pass on their genes while unfit ones don't—is a very simple idea, yet it contains great power for explaining nature and vast potential for study.

Other fields are equally driven by simple ideas. Modern geology, for example, is driven by the concept of plate tectonics, which explains the shape of the global landmasses, the rise and fall of mountain ranges, and the long-term geochemistry of our planet. Organic chemistry is driven by atomic orbital theory and the idea of hybrid orbitals, which explain the structure and reactivity of organic molecules. Molecular biology is driven by the double helix of DNA and the genetic code.

These simple ideas don't explain the details and fine fabric of natural systems, but they do provide a large structure on which more complex dynamics elaborate. A colleague of mine once said, "I have to make things simplistic enough that I can understand them." In his humble way, what he meant was that he looks for the simple explanation that captures the essence of a problem, which allows the rest of us to apply those insights to our own systems. His ability to do this is why he was elected into the U.S. National Academy of Sciences.

A simple idea, therefore, is one that finds the core of the problem. It takes no special talent to see the complex in the complex. Cutting through the clutter to see the simple in the complex is what distinguishes great scientists from the merely competent.

There are different ways to find and express a simple message. For some it would be an equation; for others, a verbal description. I have always felt that I don't understand something until I can draw a cartoon to explain it. A simple diagram or model—the clearer the picture, the better. For example, the most highly cited paper I have written was a synthesis that developed a new hypothesis about how the physical structure of soil regulates how microorganisms use nitrogen, and thus controls the nitrogen forms available to plants.[2] The essence of the paper is a cartoon illustrating these interactions among chemicals, organisms, and spatial patches in the soil (figure 3.1). It wasn't until I read Heath and Heath, though, that I realized that I was searching for the simple explanation, but being a visual person, I look for it in a picture.

A contrasting example, highlighting the difference between simple and simplistic, is another paper I published evaluating the effect of freeze-thaw cycles on microbial respiration in arctic tundra soils.[3] In some soils, freeze-thaw cycles increased respiration relative to a control, whereas in others they decreased it. Initially we didn't see any pattern as to which soils respired more versus less; that inconsistency was the simple story in the first submitted version of the paper. The reviewers, however, thought that was simplistic and said so in no uncertain terms.

2. J.P. Schimel and J. Bennett, "Nitrogen Mineralization: Challenges of a Changing Paradigm," *Ecology* 85 (2004): 591–602.

3. J.P. Schimel and J.S. Clein, "Microbial Response to Freeze-Thaw Cycles in Tundra and Taiga Soils," *Soil Biology and Biochemistry* 28 (1996): 1061–66.

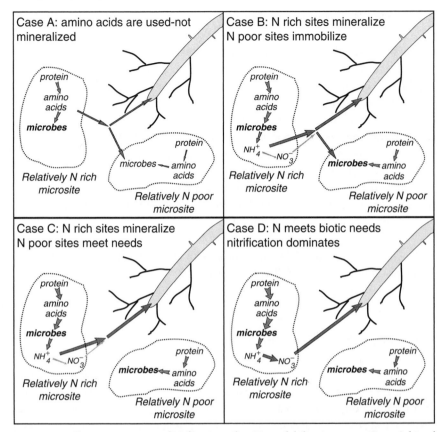

Figure 3.1. Changing patterns of N-flow in soil as N-availability increases. From Schimel and Bennett (2004).
Copyright © 2004 Ecological Society of America. Reprinted with permission.

They were right. I hadn't taken Anne Lamott's advice and listened to my characters carefully enough. We went back and banged our heads for several weeks trying to find the truly simple story in the data. Was there a coherent pattern underlying the apparent inconsistency? There was—in rich soils, freeze-thaw cycles reduced respiration, whereas in poor soils they enhanced it, a pattern that suggested possible mechanisms and insights to test in future research. It was one of those "what an idiot!" moments, where something suddenly becomes clear, and you wonder how on Earth you could have missed it before. That paper has been cited over 100 times, largely because the reviewers held our feet to the fire to do a better job of finding the simple story in the complex data. That isn't the only case where I owe reviewers thanks for criticizing me for not having done a good enough job on data analysis or story development. Of course, it's better when reviewers hang tough than when they are "nice" and let you publish less-than-perfect work. The pain of an embarrassing review lasts a few days, the pain of an embarrassing paper lasts a lifetime.

3.1.1. Simple Language: Schemas

Part of being simple is expressing your thoughts in language that builds off ideas that your readers already know. Heath and Heath borrow the term *schema* from psychology to identify ideas we bring with us to a problem. Lincoln used the images of "simple-minded soldier boy" and "wily agitator"— you can immediately flash mental pictures of those characters.

Why are schemas so important to create messages that feel simple? They are how people learn; we start with existing schemas and then attach new information to develop new, more sophisticated ones. It's hard to learn new material when you can't fit it into an existing intellectual structure—in that case, you need to build the new structure from the ground up. For example, if you were describing how alligator meat tastes, you might say:

> It's a light-colored, finely textured meat, with very little fat. It cuts easily and is moist if not overcooked. The flavor is mild.

Or you could say:

> It tastes like chicken, but a little meatier.

The first explanation describes the individual traits of alligator, but that somehow misses the point—it doesn't make it evocative. The second grounds this new idea firmly in one you probably know well: the taste of chicken. Alligator meat may not taste exactly like chicken, but this explanation gets you most of the way there.

The idea of schemas and how they relate to learning is why university science curricula are structured as they are—first-year inorganic chemistry introduces the idea of electron orbitals as energy bands that electrons can jump between. Second-year organic chemistry modifies that schema to introduce the idea of hybrid orbitals and resonance structures. Third-year physical chemistry takes this further, introducing the Schrödinger equation, which treats orbitals as probabilistic distributions of electrons. Similarly, in molecular biology we start with the simple transcription/translation model of DNA →RNA→ protein, and the idea of one gene/one product. Only after establishing those schemas do we start introducing ideas such as post-translational modification of proteins and overlapping reading frames (a single stretch of DNA may actually be part of two separate genes). Each step takes a simple schema and modifies it, making it increasingly elaborate and nuanced.

This sequential approach means that we usually start with an explanation that to an expert may seem horribly simplified or just plain wrong. A physical chemist knows that the way we explain reactions in freshman chemistry is a ghastly misrepresentation of how the systems truly work. However, you don't teach someone to swim by throwing them into the deep end of the pool and describing how to do the butterfly. You have to start simple and work up to it. You establish schemas and then expand and modify them. Building off established schemas makes ideas feel simple.

To communicate effectively in science, we need to know what schemas our audience holds so we can build from them. If we assume readers hold schemas they don't, we write above their knowledge level and confuse them, whereas if we explain schemas they do hold, they may feel that we are writing below them.

Because schemas are our core ideas, we often take them for granted. We think and write based on the schemas we and our closest colleagues hold, limiting the reach of our writing to a narrow community. Succeeding widely, however, requires reaching a broader audience, so when you use ideas and terms, stop and think about whether they relate to schemas held by the target audience. If not, don't be afraid to redefine your ideas in simpler terms and more broadly held schemas.

3.2. UNEXPECTED

Why is being unexpected important in telling a good story? Well, any paper that just presents another data set showing things we already knew, that presents a slight variation on an existing method, or that merely reinforces dogma is going to be forgettable. Most papers (even solid ones), are forgettable, because they are incremental, filling in gaps and providing additional facts that solidify a platform for launching new ideas. Incremental science can be important, but really good papers go beyond incremental to *novel*—they say something new and unexpected.

Novelty and unexpectedness lie in the questions you ask and the interpretations you develop. There are no areas of science where there aren't new questions to be asked (physicists have occasionally thought so but learned better). Few data sets don't provide the opportunity to develop new insights. Conversely, few data sets are so imbued with novelty that you can't use them to tell a boring and uninsightful story. Your job is to find what is novel and highlight the unexpected elements. Frame new questions and look for new insights. Make them clear in your writing.

In science, the key to highlighting the unexpected is through the knowledge gap theory of curiosity described by Heath and Heath. There is undoubtedly an enormous mass of knowledge on your overall topic, but your work should identify the unknown within that mass. By highlighting that unknown, identifying ignorance in the midst of knowledge, you create unexpectedness and engage a reader's curiosity.

We all work on big questions that have been around for years or decades, and we do good science by identifying new aspects of those questions—pieces that, if we accomplish them, will make progress on the bigger questions. The knowledge gaps we identify may be small, but that doesn't mean they are unimportant. Science doesn't advance by great leaps but by many small steps, each of which makes its own contribution. In any event, it is better to write about a small knowledge gap than about no knowledge gap at all.

Unfortunately, highlighting the unknown is often difficult for us. We're scientists—we know a lot, and we like to show off what we know. Particularly for junior authors, who may not be comfortable with how much they know, and how much they don't, it can feel important to show off their knowledge. But showing off knowledge doesn't create curiosity. Rather, in the words of Heath and Heath, "Our tendency is to tell people the facts. First, though, they must realize they need them." We make a good story by identifying the knowledge gap we will fill.

You frame a knowledge gap by using what is known to identify the boundaries of that knowledge. It's like framing a window—build the structure to support the area you will fill in. Identifying a knowledge gap creates curiosity. Filling that gap creates novelty.

3.3. CONCRETE

> If those who have studied the art of writing are in accord on any one point, it is this: the surest way to arouse and hold the reader's attention is by being specific, definite, and concrete.
>
> —STRUNK AND WHITE, *The Elements of Style*

As an example of the power of being concrete, I'll go back to Bill Clinton and "it's the economy, stupid." That is a concrete way of expressing a classic maxim in politics: you must stay focused. Anytime Clinton found himself being drawn into other interesting directions, the rude bluntness of "it's the economy, stupid" helped pull him back to his core message. Simple has power, but concrete adds mass to that power. A balloon is simple, but you notice more when you get hit in the head by a brick.

The importance of being concrete might seem an obvious and inherent characteristic of writing science. After all, science is about data, and data are concrete. But science is also about ideas, and ideas are abstractions—the antithesis of concrete.

Science lives with this tension between concrete data and abstract ideas. We even use the abstractions to make sense out of the concrete. The world is too complex to understand in all its detail, so we create abstractions—models and theories—to shape the complexity into structures simple enough for us to understand. In fact, being able to convert the concrete into the abstract is part of what makes someone an expert. For a novice, a specific detail is a concrete thing on its own. For an expert, it is an example of a broader set. The more we learn, the more we are able to think about a topic at a higher level of abstraction. We can get so caught up in those abstractions that it is easy to forget the concrete blocks we built them from. I struggled as a teaching assistant in introductory chemistry—I had forgotten the simple explanations my teachers had used to build concepts I took for granted, concepts like mole, valence, and stoichiometry.

Abstract and concrete, however, are not a dichotomy but a continuum, what Roy Peter Clark describes as the "Ladder of Abstraction."[4] At the top of the ladder are the widest abstractions—the simple ideas that motivate science and are broadly understandable: survival of the fittest, plate tectonics, and so on. At the bottom are the physical facts—the actual data we collect. Both of these are tractable for most readers.

The danger zone is in the middle—small-scale abstractions that are neither concrete details nor high-level schemas. This middle zone is inhabited by the concepts that are the bread and butter of scientific discourse, schemas that are typically held only by experts. Evolutionists don't spend their time discussing survival of the fittest—that is taken for granted. Rather, they write papers about sexual selection, Hardy-Weinberg equilibria, and genetic drift. Molecular biologists don't write papers about the double-helix model but about knockout mutations, ribozymes, and transcriptional silencers. When environmental engineers talk about "multimedia modeling," they don't mean audio and video but soil and water. These middle-level concepts are what outsiders consider jargon.

Scientists are drawn to the middle of the ladder of abstraction and as a result, we often write papers that are accessible to only a limited group of readers. You can't avoid the middle rungs, but you can minimize the damage—you can ground and define your specific concepts either in widely understood schemas or in the details that explain the abstractions. I discuss how to do this later in the book (particularly in chapters 11 and 14).

To illustrate the idea of grounding concepts in the concrete, consider my earlier discussion of the flow from data through information and knowledge to understanding. Would that section have made sense without the example of the discovery of the structure of DNA and the separate roles of Franklin versus Watson and Crick? By linking a concept to a concrete example, the concept itself becomes concrete—a new schema you can work with.

3.4. CREDIBLE

Science writing that isn't credible is science fiction. Credibility goes hand in hand with being concrete. We establish the credibility of our ideas by grounding them in previous work and citing those sources. We establish the credibility of our data by describing our methods, presenting the data clearly, and using appropriate statistics. We establish the credibility of our conclusions by showing that they grow from those credible data. We build a chain that extends from past work into future directions. A break anywhere in that chain makes the whole endeavour lose credibility.

I recently reviewed a proposal, and after reading the introduction, I was prepared to hate the whole thing. The ideas had potential, but instead of fleshing them out, the authors loaded them up with boldface, buzzwords, and hype. I was

---

4. R. P. Clark, *Writing Tools* (Little, Brown, 2006).

sure that with that much lipstick, the proposal had to be a pig. It wasn't concrete, and as a result it wasn't credible—the writing style undermined the content. I was surprised, however, when I got to the meat of the proposal: it was stellar. There, the authors demonstrated that their program was well thought out and would, in fact, address all the program goals. The proposal only became credible when it became concrete. That's what convinced me it was worthwhile and converted me from a skeptic to a supporter.

3.5.   EMOTIONAL

This is an awkward one for scientists. To do good science you must be dispassionate and objective about your work. There is, however, one emotion that is not only acceptable in science but fundamental to it: curiosity. We became scientists because we are curious—we are driven to solve the puzzles that nature presents. To engage us in your work, you need to engage our curiosity. You do that by asking a novel question.

If you don't ask an engaging question, and instead just offer new information, you appeal to another, weaker emotion. You appeal to our inner nerd and our love for accumulating trivia. That won't get your paper published or your proposal funded.

The E element of the SUCCES formula is thus closely aligned with U. Unexpected things create curiosity, so use that link to your benefit. You engage emotion by shifting your focus from "what *information* do I have to offer?" to "what *knowledge* to I have to offer?" Phrased differently, shift from "what's my answer?" to "what's my question?"

Working on E this way is important to enhancing the impact of a paper but it can mean life or death for a proposal. Proposals are evaluated by a panel of your peers, and your proposal is in direct competition with other good proposals. In my experience, at least twice as many proposals are considered fundable as there is money to fund. To make it from the *fundable* to the *funded* list, you need to get at least one panelist excited enough to be your advocate, arguing why your project should be funded at the expense of other good proposals. Without such an advocate, you are likely to get one of those frustrating "if we only had enough money, we would have funded you" letters. You must excite the reviewers. Excitement is the therefore the second acceptable emotion in science, and it grows from curiosity. We get excited about work that engages and then satisfies our curiosity.

3.6.   STORIES

This whole book is about telling stories—about seeing your work as a story and presenting it that way. But stories are modular; a single large story is crafted from a collection of smaller story units, threaded together. To write a good paper, you need to think about internal structure and how to integrate story modules.

For example, in chapter 2, I told a story about the role of storytelling in science. I built it from three modules, each its own story with its own characters. The first focused on Elizabeth Kolbert and her perception that scientists don't tell stories. The central characters were Kolbert, scientists, and, importantly, the idea of "story" as a character itself. In the second module, to discuss the idea that science goes from data to understanding, I used the story of the discovery of the structure of DNA. Finally, to describe how "listening to your characters" can enhance science, I used the stories of Bill Dietrich's doctoral work and that of my own. I hope that each of these short stories was sticky in its own right, and that together they created a sticky overall story.

You can use the same strategy in your writing. As you discuss your data and ideas, find units that you can package into coherent modules. Readers will be able to assimilate each piece, and it will be easier for them to see how they add up to create the whole.

These six SUCCES elements are integral to effective storytelling and science writing. Before you start writing, take the time to figure out how you are going to weave them into your work. Particularly, take the time to figure out the simple story. Build it around the key questions that will engage U and E. These will guide you in selecting the material you need to present to make the story concrete and credible.

EXERCISES

3.1. Analyze published papers

Go back to the papers you are analyzing:

Identify how the authors used each SUCCES element. Did the authors do a good job? Could they have done a better job? If so, how? Try rewriting key passages to enhance their SUCCES power.

What schemas did the authors use in building the story? Are these only held by a narrow subdiscipline or by a wider community?

3.2. Write a short article

Analyze the short articles(s) you (and your writing group colleagues) wrote for the exercise in Chapter 2.

Identify how well you and your peers used SUCCES elements. Did you do a good job? Could you have done a better job? If so, how? Rewrite key passages to enhance their SUCCES power.

4

Story Structure

All stories have a beginning, a middle, and an end.

When talking about story structure, many people will give you this simple platitude. Of course, if the proverbial monkey at a typewriter pounds away, the gibber-
ish it types will start, end, and have stuff in the middle. Is this classic saying just a
meaningless reflection of an obvious physical necessity? In fact, no. Beginning,
middle, and end, however, don't reflect physical positions but story elements that
carry out specific functions.

All stories have common elements that are necessary to make them engaging
and memorable. To respond to their specific pressures, however, different genres
have evolved traditions for assembling the elements, producing a number of standard structures. In science, we have adapted several of these for solving writing
problems, but most of us are probably unaware of which ones we are using and
why. Understanding the common elements, the ways you can put them together,
and when each structure works provides a powerful tool for approaching different
writing challenges.

There are four elements that underlie the structure of all stories, including
those we write in science:

Opening (O): Whom is the story about? Who are the characters? Where
does it take place? What do you need to understand about the
situation to follow the story? What is the larger problem you
are addressing?

Challenge (C): What do your characters need to accomplish? What specific question do you propose to answer?

Action (A): What happens to address the challenge? In a paper, this describes the work you did; in a proposal, it describes the work you hope to do.

Resolution (R): How have the characters and their world changed as a result of the action? This is your conclusion—what did you learn from your work?

Together these elements generate the acronym OCAR, a concept that echoes throughout this book, whether we are talking about whole papers, sections, paragraphs, or even individual sentences. Understanding how to manage the OCAR elements is at the heart of successful writing. A story lacking any element from it will be unsatisfying, ineffective, and slippery, rather than sticky.

4.1. THE FOUR CORE STORY STRUCTURES

There are four story structures that different genres use regularly. Which one to use depends largely on the audience's patience. Are readers willing to wait to get the point of the story, or do they want to see it right away? In science, the audience's patience varies with what you are writing—a paper for a specialist journal, a paper for a generalist journal, a proposal, or a piece for the public. The following structures span from targeting the most to the least patient of audiences.

OCAR Structure: The simplest, but also the most slowly developing structure is to simply take the OCAR elements in sequence. This is used in some fiction; books may take several chapters to introduce the characters before defining what those characters must do. My favorite example of this is J. R. R. Tolkien's *The Lord of the Rings*. Only in chapter 2 do we get the first inkling of Frodo and Sam's challenge—to take the Ring of Power to Mt. Doom to destroy it. It takes until halfway through the first book to learn the full challenge—take the ring to the fire and restore Aragorn to the throne of Gondor.

OCAR is the structure we use most frequently in science because readers are patient—they want to assess your ideas and results as they develop. They want to see the evidence and the arguments clearly laid out before any conclusions are presented. Thus, a paper's challenge is presented at the end of the introduction, and its conclusion comes at end. Because OCAR is slow to develop, it requires a patient audience, one that is willing to watch the story unfold.

ABDCE Structure: Not all audiences are patient enough for OCAR. For those a step less patient, a structure known as ABDCE works well. This is the structure that modern fiction writers and scientific proposal writers use most frequently:

Action (A): Start with a dramatic action to immediately engage readers and entice them to keep reading.

Background (B): Fill the readers in on the characters and setting so they can understand the story as it develops.

Development (D): Follow the action as the story develops to the climax.

Climax (C): Bring all the threads of the story together and address them.

Ending (E): What happened to the characters after the climax? (This is the same as the resolution.)

The difference between ABDCE and OCAR is that ABDCE front-loads the story by moving the challenge up and collapsing it into the opening to create the initial "action"—an exciting start to grab your attention. Thus, A and B together comprise the O and C elements.

ABDCE gets the reader into the story faster by launching directly into the challenge, so it is good with an impatient audience, such as proposal reviewers. But it is less efficient than OCAR in moving the story forward—after the initial action, you have to back up and fill in the background. That inefficiency is a fair trade, however, if it gets readers committed to the rest of the story.

ABDCE reaches its apex in mystery and adventure stories. James Bond movies, for example, always start with wild action sequences—boat chases, gunfights, explosions. Or consider the beginning of *C Is for Corpse*, by Sue Grafton:

> I met Bobby Callahan on Monday of that week. By Thursday, he was dead. He was convinced someone was trying to kill him and it turned out to be true, but none of us figured it out in time to save him. I've never worked for a dead man before and I hope I won't have to do it again. This report is for him, for whatever it's worth.
>
> My name is Kinsey Millhone. I'm a licensed private investigator.

Here we are thrown immediately into the action and introduced to the key characters, in this case the dead Bobby Callahan and the living Kinsey Millhone. We are also given the challenge that Kinsey faces: who killed Callahan, and why?

This action-first structure has become common in a publishing world where overloaded editors may judge whole novels by the first page, but it isn't new. Here is the opening from a well-known story: "All the survivors of the war had reached their homes by now and so put the perils of battle and the sea behind them. Odysseus alone was prevented from returning to the home and wife he yearned for by that powerful goddess, the Nymph Calypso, who longed for him to marry her, and kept him in her vaulted cave." This is the opening to *The Odyssey* by Homer, one of the oldest recorded stories in human history. The classical in medias res structure of the epic is straight ABDCE.

One aspect to both the OCAR and ABDCE structures is that they have a resolution that shows how overcoming the challenge has changed the characters and their world. While people often see the climax as the focal point of the story—the point that all the action builds to and is the most exciting part of the story—ultimately, the resolution makes sense out of that action. The resolution wraps up the story that was introduced in the opening; it closes the circle.

"The climax is that major event, usually toward the end, that brings all the tunes you have been playing so far into one major chord, after which at least one of your people is profoundly changed. If someone isn't changed, then what is the point of your story?".

ANNE LAMOTT, *Bird by Bird*

We want and need to know how our characters have changed for a story to be satisfying. Cinderella has to marry Prince Charming. The hero has to overcome his personal demons and move on to a different life. In science, we have to see how our understanding of the world has changed.

The importance of closing the loop means that a good story is circular; at the end, it must come back to the beginning. However, because things have changed, the "beginning" has moved. Thus, a story isn't truly a circle, but a spiral (see figure 4.1). Highlighting this spiral structure is key to making an OCAR or ABDCE story powerful. A story without a resolution, such as Samuel Beckett's *Waiting for Godot*, falls into the "Theater of the Absurd." A science paper without a resolution falls into the reject bin.

LD Structure: ABDCE front-loads the story more than OCAR by collapsing the challenge into the opening, but some audiences are so impatient they won't stick around for a resolution. For them, you need to intensify the front-loading. The most extreme case of this is used by newspaper reporters. Reporters use a structure that they call the "inverted pyramid"; I call it Lead/Development (LD) to highlight its key functional elements. In LD structure, the core of the story is in the first sentences (the lead, L) and the rest fills out and develops the story (the development, D). In LD structure, the lead collapses the opening, challenge, *and* resolution into a single short section, possibly as little as a single sentence.

The most important sentence in any article is the first one. If it doesn't induce the reader to proceed to the second sentence, your article is dead. And if the second sentence doesn't induce him to continue to the third sentence, it's equally dead. Of such a progression of sentences, each tugging the reader forward until he is hooked, a writer constructs that fateful unit, the "lead."

WILLIAM ZINSSER, *On Writing Well*

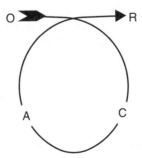

Figure 4.1. How an OCAR story makes a spiral: the story comes back to its starting point, but that point has moved.

LD structure developed in the newspaper business under the lash of the physical restrictions on space. A newspaper's front page has five or more stories that
start there but may finish in the bowels of section B. Many people read only page
1; they don't bother opening the paper to finish stories. Additionally, in finishing
layout on a tight deadline, an editor may slash the last few paragraphs of a story to
squeeze it in. So writers cannot put the point at the end of the story—if they did,
readers might never see it. It *must* go at the beginning.

LDR Structure: Journalists who write for magazines suffer a similar challenge
to those who write for newspapers, although less extreme. There are many articles
in a magazine, but they don't all start together on page 1, and you can't just skim
the first page to get the sense of everything in the issue. Each story must keep you
turning the pages to see all the *advertisements*. So the lead must be engaging
enough to commit you to the story, but the writer can realistically hope that
if you start reading, you may actually finish. Also, magazines aren't under the
same production pressure as newspapers—editors don't cut paragraphs to make
space. Magazine writers can afford to worry about ending well with an effective
resolution.

> "Often it takes just a few sentences to wrap things up. Ideally they should
> encapsulate the idea of the piece and conclude with a sentence that jolts us
> with its fitness or unexpectedness".
>
> WILLIAM ZINSSER, *On Writing Well*

Thus, magazine journalists use a story structure I describe as Lead/
Development/Resolution (LDR). It's conceptually similar to the newspaper LD
but tuned to a slightly more patient audience.

These structures give us a continuum, based on readers' patience:

OCAR:	*Slowest—take your time working into the story.*
ABDCE:	*Faster—get right into the action.*
LDR:	*Faster yet—but people will read to the end.*
LD:	*Fastest—the whole story is up front.*

These story structures define the information you must present in different
places. Readers intuitively understand and respond to different structures; they
know how to identify the critical locations, and they anticipate the information
that will appear. The opening is the first short section, and we use certain signals
to tell the reader that we have reached the challenge or resolution. Readers take
whatever information you put at those key locations—O, C, and R—and accept it
as your opening, challenge, and resolution. If you put the wrong information
there, they will get the wrong message.

You should be able to read the O, C, and R of a paper, and still get its key points.
If you know the problem, the specific questions, the general approach to answer-
ing them, and the conclusions, you may have gotten all you need from that paper.
You'll certainly know whether you need to go back and read it fully.

4.2. APPLYING STORY STRUCTURE TO SCIENCE WRITING

The OCAR functions are as central to a scientific paper as they are to a work of fiction. A good paper or proposal describes the larger problem and central "characters" (O); it frames an interesting question (C); it presents your research plan and results, developing the action (A); and it leaves the reader with an important conclusion about how our understanding of the world has changed as a result of the work (R).

Different types of science writing, however, use different structures to achieve best results. Papers for specialist journals generally use straight OCAR, framing the challenge at the end of the introduction. Readers of these journals are patient. They want your information and your thoughts, and they want to evaluate the progression from ideas through results to conclusions. In fact, in these journals, a lead-based paper might be considered suspect. Front-loading the story with conclusions could make it seem like you knew the story you wanted to tell and were simply forcing the data into that story—that is, you are trying to prove rather than test your ideas—a no-no going all the way back to the foundations of the philosophy of science.

In contrast, generalist journals, such as *Nature* or *Science*, need a faster structure, often closer to LDR. The need arises because the greatest hurdle to getting published in these journals is the editor. *Nature* and *Science* editors are professional editors, rather than practicing scientists who serve as editors on the side (as is typical for specialist journals); they are generalists, and they get swamped with manuscripts. They have to perform triage, deciding quickly whether a paper seems novel and important enough to invest the time to send it out for review. (I have submitted papers to *Nature* on Friday afternoon and had the rejection first thing Monday morning!) To make that first cut and get your paper sent out for review, you need a good lead.

A famous use of a lead-based structure in science is Francis Watson and James Crick's famous paper "Molecular Structure of Nucleic Acids: A Structure for Deoxyribose Nucleic Acid," in which the opening was: "We wish to suggest a structure for the salt of deoxyribose nucleic acid (D.N.A.). This structure has novel features which are of considerable biological interest."[1] The rest of the paper describes this structure and resolves with the statement, "It has not escaped our notice that the specific pairing we have postulated immediately suggests a possible copying mechanism for the genetic material." They didn't bother to elaborate on that mechanism—it was obvious enough that they didn't need to.

Moving on to the least patient audience of all, we come to proposals. We review proposals because we owe it to the agencies that fund our work. We review proposals on airplanes when we would rather read a novel, watch a movie, or sleep. Patient? No. A proposal must convince reviewers that the topic identified in the

---

1. J. D. Watson and F. H. C. Crick, "Molecular Structure of Nucleic Acids—A Structure for Deoxyribose Nucleic Acid," *Nature* 171 (1953): 737–38.

opening is important and then compel them with the excitement of the questions posed in the challenge. If it fails to do this, it is dead.

When I review proposals I make a "no/maybe" cut by the end of the introduction, and if it's a "no," that is irrevocable. I then only read the rest to be able to give feedback on how to improve the proposal for resubmission. A good experimental design can never compensate for boring questions. A "maybe" at that first cut means the questions are exciting, in which case I read the rest to see whether the experimental design is adequate to answer them. A saying I once heard attributed to D. A. Crossley at the University of Georgia said, "If you haven't told them in the first two pages, you haven't told them." To "tell them" in the first two pages requires a front-loaded structure: either ABDCE or LDR.

4.3. MAPPING OCAR ONTO IMRaD

I have argued that most scientific papers follow an OCAR structure, but you won't find a paper with sections labeled Opening, Challenge, Action, and Resolution. Instead, we usually write papers using some variation of IMRaD: Introduction, Methods, Results, and Discussion. So why discuss OCAR instead of IMRaD? First, even when the physical sections of the paper follow IMRaD, the conceptual structure of the story generally follows OCAR. Second, there are many permutations of IMRaD—some fields routinely combine Results and Discussion and some integrate Methods as well. Third, some types of writing—notably review papers and proposals—don't follow IMRaD at all. Regardless of a piece's form, however, it must still cover the OCAR bases. While IMRaD is a rule, OCAR is a principle.

Because IMRaD is the most common physical structure for science papers, I briefly discuss how OCAR maps onto IMRaD.

Introduction: This has three subsections, although they are rarely broken out
as such:

Opening: This is typically the first paragraph that introduces the larger problem the paper is targeting. What is the context, and what are the characters we are studying?

Background: What information does the reader need to understand the specific work the authors did, why it is important, and what it will contribute to the larger issue? I consider this an extension of the O, as it fleshes out introducing the characters.

Challenge: What are the specific hypotheses/questions/goals of the current work?

Materials and Methods: This begins describing the action—what did you do?
Results: This continues the action by describing your findings.

Discussion: This develops to the climax and the resolution. What did it all mean, and what have you learned? It often ends with a conclusions subsection that is the resolution.

Thus, opening and challenge block the beginning and the end of the Introduction. Action encompasses the M&M, Results, and much of the Discussion. The resolution is the last section of the Discussion.

The resolution is as important in science as in fiction. As Anne Lamott points out, for fiction, at the end, we want to know that "one of your people is profoundly changed." For science, at the end, we want to know how our understanding of the world has changed as a result of your work, or in the case of a proposal, how you think our understanding of the world *will* change. The resolution must map back to the opening. It must say something about the larger problem you identified there.

A closely related aspect of mapping OCAR onto IMRaD is that scientific papers have an hourglass shape to their content (see figure 4.2). They open with a problem of wide interest. Then they narrow down to a set of very specific questions that contribute to understanding that larger issue. Those specific questions comprise the challenge and define the width of the "neck" of the hourglass—that is as narrow as it is going to get. The Methods and Results sections stay narrow. Finally, as results are discussed, the context of the story expands toward conclusions that are more general and connect back to the problem developed in the opening. Importantly, the conclusions should address a topic as "wide" as the opening. That is what makes the story circular—in the resolution, you come back to the issue targeted in the opening.

In the following chapters, I discuss each of these elements in a science story: O, C, A, and R. How do we write them to most powerfully develop the stories we are telling and so create the most compelling papers and proposals?

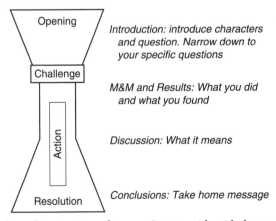

Figure 4.2. The hourglass structure of a paper. It starts wide with the opening, narrows with the challenge and action, and widens back out again at the resolution.

EXERCISES

4.1. Analyze published papers

    A. Evaluate the papers you are reading. Which story structure do they use? Where are the OCAR elements? Are they effective?
    B. Pick a proposal. Analyze it in the same way.
    C. Go back to chapter 2 and the exercise I asked you to do to start writing a short piece. Look at those questions. Notice that I was asking you to define your OCAR elements.

4.2. Write a short article

Go back and evaluate your short article. Repeat the analysis you did on the pieces above. What story structure did you use? Was it appropriate for your intended audience? Did you have the OCAR elements in place?

    If you (or your writing group members) feel that you haven't effectively developed the OCAR elements, go back and rewrite the piece to more powerfully emphasize those elements. Make sure that your opening, challenge, and resolution sections are clear and effective.

5

The Opening

The most important sentence in any article is the first one.

—WILLIAM ZINSSER, *On Writing Well*

Initial impressions are strong and lasting. Your first words have great leverage, making the beginning of a paper a "power position." You must use that power to accomplish three goals: identify the problem that drives the research, introduce the characters, and target an audience. If you're clever, you can foreshadow the challenge and even the conclusions. By establishing the paper's focus and tone, the opening identifies your intended audience—whom do you want to read your work and how do you want them to think about it?

You must start well. Your first sentences get readers moving and set the direction; you establish their expectations and generate momentum. If you start in one direction and then abruptly switch, readers get mental whiplash as they try to follow. Potentially worse, if the opening is unclear and doesn't go in *any* direction, they will sit twiddling their thumbs, waiting to figure out where to go.

The opening begins with a single sentence but typically encompasses the first paragraph, and sometimes several more. In a short paper or one for a narrow audience of experts, you can quickly remind people of a problem they already know. When you target a broader audience, one made up of people who hold different schemas than you do, you may need a longer and more complex opening.

5.1. EXAMPLES OF GOOD OPENINGS

Here are openings from different areas of science. The fundamental question for each is: does it achieve the three goals? Is it clear what the paper is about? Does it frame the problem? Does it introduce the critical characters? Read the openings and answer these questions, before going on to my analysis. Do you agree with my assessment?

This first example is from a synthesis paper I wrote reevaluating our understanding of how nitrogen (N) is processed in soil.

Example 5.1
Since the late 1800s, N mineralization has been the perceived center point of the soil N cycle and the process that controls N availability to plants.[1]

The key word in this sentence is *perceived*, a distinctive and unusual word that draws your attention. Clearly, this paper is going to challenge that perception. Additionally, there is going to be a historical element—evaluating how the perception has changed since the late 1800s.

The second example is from a study evaluating whether giving pregnant women supplemental folic acid may cause their children to develop asthma.

Example 5.2
Current public health guidelines in the United States, the United Kingdom, and Australia recommend that women consume a supplemental dose of 400 µg of folic acid per day in the month preceding and during the first trimester of pregnancy to reduce the risk of neural tube defects in children.[2]

Can you imagine that this paper is *not* going to challenge that 400 µg recommendation? That sentence doesn't give the grounds for challenging it, but because the title of the article highlights childhood asthma, you can infer the entire story: folic acid supplements during pregnancy may increase the risk of childhood asthma.

In these first two examples, the opening sentences are dramatic and launch quickly into the story. Frequently, however, openings require several steps to develop the issue, as illustrated in a paper on geomorphology that analyzed how

1. J. P. Schimel and J. Bennett, "Nitrogen Mineralization: Challenges of a Changing Paradigm," *Ecology* 85 (2004): 591–602.

2. M. J. Whitrow, V. M. Moore, A. R. Rumbold, and M. J. Davies, "Effect of Supplemental Folic Acid in Pregnancy on Childhood Asthma: A Prospective Birth Cohort Study," *American Journal of Epidemiology* (2009).

the size of the sediment particles created during erosion affect their abrasive properties and how fast they cut a river channel.

Example 5.3

The topography of mountainous landscapes is created by the interaction of rock uplift and erosion. River incision into bedrock is the key erosional process that controls the rate of landscape response to changes in rock uplift rate and climate.[3]

Clearly this paper is going to evaluate river incision ("the key erosional process"), but developing that point required two sentences. The first frames the focus of the story: the topography of mountainous landscapes, something many people find interesting. It also introduces the key character of erosion. The second sentence picks up the idea of erosion and develops a specific focus: river incision. It would have been hard to get all that into a single sentence, so the authors used an initial "positioning sentence" from which they could launch to make their specific point.

Sometimes the opening needs to be longer and can include the entire first paragraph. Example 5.4 is from materials chemistry and explores how the molecular structure of organic polymers affects their potential as semiconductors. In this example, I include the first and last sentence of the opening paragraph.

Example 5.4

Conjugated polymers are novel materials that combine the optoelectronic properties of semiconductors with the mechanical properties and processing advantages of plastics. . . . Thus, conjugated polymers offer the possibility for use in devices such as plastic LEDs, photovoltaics, transistors, and in completely new applications such as flexible displays.[4]

The first sentence frames the overall topic of the story—conjugated polymers are going to be exciting new materials for developing plastic optoelectronic materials. That is highlighted and made concrete in the last sentence, which describes devices that might be made. Whereas the opening sentence deals in abstractions, the last one sets the story in concrete terms—real applications. From this, we can infer that the rest of the paper is going to be about how to perfect the polymers so that they can be used to produce these devices.

3. L. S. Sklar and W. E. Dietrich, "Sediment and Rock Strength Controls on River Incision into Bedrock," *Geology* 29 (2001): 1087–90.

4. B. J. Schwartz, "Conjugated Polymers as Molecular Materials: How Chain Conformation and Film Morphology Influence Energy Transfer and Interchain Interactions," *Annual Review of Physical Chemistry* 54 (2003) :141–72.

5.2. BAD OPENINGS

The foregoing were all examples of effective openings. They vary in length, but all identify a problem of broad interest and give the reader a sense of where the story is going. But they raise the obvious question: how do you write an *ineffective* opening? Since the opening is supposed to provide direction about where the story is going, there are two obvious ways to fail: provide either misdirection or no direction.

5.2.1. Misdirection

An example of misdirection comes from a paper I wrote; this isn't terrible, but it could have been better. I analyzed the processes that control how much methane (CH_4, an important greenhouse gas) is released from tundra soils of the Alaskan Arctic. Bacteria known as methanogens produce the CH_4, but then plants transport it out of the soil through their roots. Here is the first paragraph and the first sentence of the second paragraph.

> Example 5.5
> Plants are a critical control of CH_4 dynamics in wetland ecosystems. They supply C [carbon] to the soil methanogenic community both through production of soil organic matter, and as fresh exudates and residues. Fresh plant material may be an important CH_4 precursor even in an organic matter–rich peat soil. Strong correlations between net primary productivity and system-level CH_4 fluxes across a wide range of ecosystems highlight the importance of plant C inputs.
>
> Vascular plants, however, also transport CH_4 out of soil and sediment, effectively bypassing the aerobic zone of CH_4 oxidation.[5]

Why is this misdirection? The first paragraph introduces plants as the central character of the story and argues that they control CH_4 fluxes. That is accurate. But the first paragraph develops a story about how plants control CH_4 by feeding carbon to methanogens, and a reader would likely assume that is what the whole paper is about. That is inaccurate, which you realize once you begin the second paragraph. It opens by introducing a new mechanism—plants transport CH_4 out of the soil. That cuts the readers adrift and leaves them momentarily wondering: was the opening paragraph a false lead, highlighting what we thought the mechanism was, but that I will contradict—so the paper is going to be about transport? Or am I introducing an additional mechanism— so the paper will be about both? I started in one direction, but then struck out in a different one; that is misdirection, and it's confusing.

5. J. P. Schimel, "Plant Transport and Methane Production as Controls on Methane Flux from Arctic Wet Meadow Tundra," *Biogeochemistry* 28 (1995): 183–200.

I could have written this better and avoided any potential confusion by changing the first sentence, making it a broader positioning statement. "Plants control CH_4 dynamics in wetland ecosystems by two mechanisms. The first is to supply C to the soil methanogenic community . . ." This would have let you know that the paper is about both mechanisms and might imply that it evaluates the balance between them. It signals that the next sentences or paragraphs will identify and describe those mechanisms. Even though the first paragraph is about substrate supply, you would know that there is more coming, so the second paragraph would not feel like it was changing direction but completing the direction I had started.

Fixing this opening involved an almost trivial change, but it would have made the reader's job easier. Unfortunately you can't go back and rewrite a published paper; all you can do is try to learn from mistakes (in this case, mine) and make the next paper better.

5.2.2. No Direction

The other common error in the opening is giving no direction. Consider the following example.

Example 5.6

In meiosis, genes that are always transmitted together are described as showing "linkage." Linkage, however, can be incomplete, due to the exchange of segments of DNA when chromosomes are paired. This incomplete linkage can lead to the creation of new pairings of alleles, creating new lineages with distinct sets of traits.

Is this paper about the evolution of sex chromosomes in guppies, the distribution of Tay-Sachs disease among Louisiana Cajuns, or the ecology of the potato blight fungus *Phytophthora infestans*? You can't tell—this opening offers no direction as to where the story is going. Rather, it goes over basic, textbook material about eukaryotic genetics that should be second nature to most readers. It explains a schema that scholars in this field don't need explained.

Using an opening that explains a widely held schema is a flaw common with inexperienced writers. Developing scholars are still learning the material and assimilating it into their schemas. It isn't yet ingrained knowledge, and the process of laying out the information and arguments, step by step, is part of what ingrains it to form the schema. Many developing scholars, therefore, have a hard time jumping over this material by assuming that their readers take it for granted. Rather, they are collecting their own thoughts and putting them down.

There is nothing wrong with explaining things for yourself in a first draft. Many authors aren't sure where they are going when they start, and it is not until the second or third paragraph that they get into the meat of the story. If you do this, though, when you revise, figure out where the real story starts and delete everything before that. At a writers' conference my wife attended, a well-known author said that he sometimes has to delete several chapters to get to where the story begins.

5.3. TARGETING YOUR AUDIENCE

The way you introduce your problem and your characters affects the audience's attitude toward the work and maybe whether they continue reading. You must know the intended audience to tailor the writing to them. This is particularly important for generalist papers and proposals, where reviewers and readers are impatient and may not be familiar with the schemas of your discipline. In such cases, the opening may determine the success of the entire piece—if it is published or funded.

Consider two papers about bacteria in the ocean; one was for a specialist and the other for a generalist journal. They both address the standard message in microbial ecology that we have identified and grown in culture less than 1 percent of all the bacteria that exist.

Example 5.7
For a specialist journal Epifluorescence microscopy and direct viable counting methods have shown that only 0.01 to 0.1% of all the microbial cells from marine environments form colonies on standard agar plates. Much of the discrepancy between direct counts and plate counts has been explained by measurements of microbial diversity that employed 16S rRNA gene sequencing without cultivation. The present consensus is that many of the most abundant marine microbial groups are not yet cultivated.[6]

This opening makes an important point that is fundamental to the story— organisms have not been cultured on standard agar—and so foreshadows that the authors will grow new organisms on nonstandard agar. The characters are methods (for counting and culturing) and microbes in the sea, characters that environmental microbiologists identify with and care about. As an alternative, however, consider the following.

Example 5.8
For a generalist journal: Antonie van Leeuwenhoek (1632–1723), the first observer of bacteria, would be surprised that over 99% of microbes in the sea

6. J. C. Cho and S. J. Giovannoni, "Cultivation and Growth Characteristics of a Diverse Group of Oligotrophic Marine Gammaproteobacteria," *Applied and Environmental Microbiology* 70 (2004): 432–40.

remained unseen until after Viking Lander (1976) set out to seek microbial life on Mars.[7]

This opening says something similar to example 5.7, but these authors (Farooq Azam and Alexandra Worden) were targeting the editors and readers of *Science*, a group with limited interest in methods for culturing bacteria. So, they opened with a short story whose characters are Antonie van Leeuwenhoek, the Viking Lander, and microbes. Most scientists have probably heard of van Leeuwenhoek and his "wee animalcules," and of the Viking Lander, so this speaks to a wide readership in a way that culturing bacteria cannot. To make their story engaging, Azam and Worden pulled strongly on the SUCCES elements. It is simple, and it is unexpected—we are searching Mars for life when we haven't found 99 percent or more of the life on this planet. It is concrete and credible, backed up by specifics. It is also emotional, pulling on your curiosity and amazement—we've been at this for 300 years and have seen at best 1 percent of the bacteria that exist!? This is a powerful start for a *Science* paper. I'm not surprised it was published there, the science was excellent—and so was the storytelling.

The opening that Azam and Worden use, however, might make the readers of a specialist microbial ecology journal uncomfortable. It launches the story so flashily that it would stick out. More important, it makes a point that most microbial ecologists already know—this isn't their knowledge gap.

While targeting the right audience is important in papers, it can be life or death in proposals. As an example of this power, consider a project I was part of in which we studied coastal redwood forests in California. These forests are a treasure, but they exist in a region where it often doesn't rain from April to November. During the long, dry summers, the trees depend on fog for water. Climate change may alter the amount and timing of fog, potentially placing those forests at risk. But there are other foggy forests, so this represents an important and general ecological phenomenon.

We submitted similar proposals to two agencies: the National Science Foundation's (NSF) Ecosystem Science Program, and another that had a management focus (for this exercise, think California Environmental Protection Agency; CalEPA). We used different openings.

Example 5.9:
The influence of fog on ecological and hydrological processes in coastal zones has long intrigued scientists.

Example 5.10:
California's coastal forests are among its most distinctive and treasured natural resources.

---

7. F. Azam and A. Z. Worden, "Microbes, Molecules, and Marine Ecosystems," *Science* 303 (2004): 1622–24.

Imagine submitting a proposal with the first sentence to the NSF—a reviewer might well be drawn in on the idea: "I never thought about fog being that important in ecology; I should read further." A CalEPA reviewer's response, on the other hand, would probably be closer to: "Ivory Tower academe that is irrelvant to my mission; I can ignore this one."

The responses would differ with the second example. An NSF reviewer would likely think: "Regional interest and an environmental protection focus—this isn't NSF science; I can ignore this one." The CalEPA reviewer, though, might think: "That's true, coastal forests are important resources that we *are* responsible for protecting. I'd better read further to find out how this research may help me do that."

How we framed the problem here was critical. An effective first sentence might open the door to funding. An ineffective one could close it.

5.4. OPENING FOR A BROADER AUDIENCE: THE TWO-STEP OPENING

When you target experts in your field, you can open quickly, building off the discipline's core schemas. Sometimes, though, you need to target a broader audience—people who might be interested in your work but don't necessarily hold the same schemas you do. To do this, you need to open with an issue that engages your target audience, but then modulate it to one you want to work with. That requires a multistep opening in which you take time to introduce and then redefine the focus.

An example of this two-step approach is a paper written by Mike Weintraub, a doctoral student of mine. The paper described a laboratory experiment evaluating the factors that control decomposition of the organic material that makes up arctic tundra soils. Though the work was narrow, the opening was wide.

Example 5.11:
The Arctic has become a focus of attention because global warming is expected to be the most severe at extreme latitudes. The thick organic soils of the tundra contain large stocks of carbon (C), and these soils may act as either a source or a sink for atmospheric carbon dioxide (CO_2). It has been suggested that as the climate warms, increased organic matter decomposition will release CO_2 to the atmosphere, contributing to warming and creating a positive feedback that results in further increases in atmospheric CO_2. Alternatively, it has been argued that increased decomposition will release bound nitrogen (N) and other nutrients in the soil and thereby enhance plant growth, since plant growth is nutrient-limited in arctic tundra. Increased plant growth would allow the tundra to be a sink for atmospheric C because plant material has a wider C/N ratio than soil organic matter. Thus, the direction the C balance of the arctic will shift with warming is unclear and depends

on interactions between soil C and N cycling that we still do not understand in the tundra.[8]

Weintraub opened the paper by discussing the importance of tundra soils in the global carbon cycle and then worked down in scale through several issues that regulate tundra soil carbon. Only at the end of the paragraph did he frame the specific issue: the interactions between soil C and N cycling. He was writing for an audience of global change scientists, trying to convince them that this paper was something they should read, rather than targeting tundra soil ecologists, of which there are about a dozen worldwide. One result of the broad way he framed the story was that he won the 2003 Arctic Consortium of the United States award for "Best Student Paper in Interdisciplinary Arctic Science." Weintraub was able to structure the story so that it spoke to an interdisciplinary audience. The award was a result of effective storytelling, rather than inherently interdisciplinary measurements.

In proposals, quickly engaging the reviewers is critical, so you may need to use this two-step approach. Review panel members come from diverse subfields and may not be expert in your specific topic. Writing the proposal for the panel means framing the issues broadly, in concerns held by most members. From there, you can narrow in on the specific research you propose.

I used this two-step strategy in a proposal I wrote to study plant succession—how plant communities colonize a new site and then are replaced by a series of new communities over time. I wanted to study succession in floodplain forests in the interior of Alaska, specifically how tannins produced by balsam poplar trees affect the soil microorganisms that regulate nutrient availability, and thus make the environment more favorable for poplar. I opened the proposal with the following.

Example 5.12
Succession has been a central theme in ecological research for almost a hundred years. Two questions have directed much of that research:

What causes the shifts in communities?
How do ecological processes change as a result of these community shifts?

These questions are linked through a feedback loop: plants affect soil processes which in turn affect plant community structure.

Although soil microbes and the processes they carry out were the central characters in my story, I did not introduce them in the first sentence or even the first few sentences. I did that deliberately—I submitted the proposal to the ecology program, and I knew the reviewers were likely to be plant (rather than microbial) ecologists.

8. M. N. Weintraub and J. P. Schimel, "Interactions between Carbon and Nitrogen Mineralization and Soil Organic Matter Chemistry in Arctic Tundra Soils," *Ecosystems* 6 (2003): 129–43.

I wanted to engage them with a topic they were interested in (plant succession and the factors that regulate it), and then transition to the specific topic that I was going to develop (tannin effects on soil microbes), building the connection between their interests and my work.

I call this a two-step opening for two reasons. One is to highlight that it does take two steps, but also to highlight that, like the dance, it is must be quick—if you take more than two steps, you will stumble.

5.5. CHANGING STYLE FOR DIFFERENT AUDIENCES

It is a principle of effective communication that you need to adapt your language, style, and approach to deal with different media and different audiences. To highlight how a skilled writer does this, consider this opening from another paper by Azam.

> Example 5.13:
> Larry Pomeroy's seminal paper revolutionized our concepts of the ocean's food web by proposing that microorganisms mediate a large fraction of the energy flow in pelagic marine ecosystems. Before 1974, bacteria and protozoa were not included as significant components of food web models. Pomeroy argued forcefully that heterotrophic microorganisms, the "unseen strands in the ocean's food web," must be incorporated into ecosystem models.[9]

In contrast to example 5.8, this uses more technical language and targets an audience of marine ecologists. You wouldn't write "microbes mediate a large fraction of the energy flow in pelagic marine ecosystems" if you wanted a physicist to read it—they might not know what a "pelagic marine ecosystem" is or have the schema of how energy flows through the marine food web to pick up the implications. Despite that, this opening has a clear dynamic voice—it is easy and engaging to read without seeming the least bit unprofessional. That is the product of someone who is good with both language and storytelling. This is a strong opening that effectively engages SUCCES elements. It is concrete, giving dates and directly attributing Pomeroy's paper. It is emotional, pulling on words like *revolutionary* and *argued forcefully* to create a sense of conflict. It even draws on the U factor by setting up the contrast between the thinking before and after 1974. This opening frames the story—it is going to be about the role of microbes in the ocean's food web and how our understanding of it has changed. The authors introduce key characters—Pomeroy's paper and marine food webs—and so establish the starting point.

9. F. Azam, D. C. Smith, G. F. Steward, and A. Hagström, "Bacteria-Organic Matter Coupling and Its Significance for Oceanic Carbon Cycling," *Microbial Ecology* 28 (1994): 167–79.

Whereas the *Science* paper targeted a wide audience, this one aimed more narrowly—it starts by naming Pomeroy's seminal paper and so constrains the target audience to people who already know that paper. Where the opening to the *Science* article might put off readers of a technical journal, this one is written to engage them. The broader readership of *Science* is free to read this paper, but they aren't actively courted. In fact, they may be subtly discouraged from coming to this party; "Larry Pomeroy's seminal paper" is the secret password to get in.

Skilled writers know their audiences and think carefully about what works for them. As you gain experience, these choices become easier and require less conscious effort. To gain that experience, analyze what works, what doesn't, and your own decisions—who is your audience and what are their schemas? Could you write in a way that would expand that audience? The opening is critical to that answer—do you want to target people who know Pomeroy's paper, or everyone who has ever read about van Leeuwenhoek's work? Let your opening signal those choices.

5.6. HOW WIDE SHOULD YOUR OPENING BE?

How widely you should cast your net with the opening? Remember—getting published is not the ultimate goal; getting *cited* is. You want people to use your work. Ideally, therefore, you would like it to be read and valued by a wide community. So you should set your opening, the top of the hourglass, to draw in as broad a readership as you can manage. The opening tells readers what the story is about and establishes a compact with them. You must deliver on that compact. To achieve that, the bottom of the hourglass should be the same width as the top (figure 5.1a). If you cast your opening too widely and the top of the hourglass is wider than the bottom, readers will feel cheated (figure 5.1b). Consider the following opening and two potential resolutions.

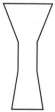

A. Opening wider than resolution: overpromising. Your readers will feel cheated.

B. On target. Your readers will be satisfied.

C. Resolution wider than opening: underpromising. Your readers won't ever see that you are telling a story that would interest them.

Figure 5.1. Matching the opening to the resolution.

Example 5.14:

> *Opening:* The Arctic is important in the global climate system because tundra soils store a large amount of carbon that may be released to the atmosphere as CO_2. An important recent discovery is that wintertime CO_2 fluxes from soil are large.
>
> **Resolution 1:** Developing a reliable model of CO_2 fluxes in the Arctic therefore requires a better model of winter C cycling processes.
>
> **Resolution 2:** In the arctic tundra, microbial community composition changes little through the winter.

Resolution 1 is framed at roughly the same "width" as the opening. The opening said the story was about wintertime tundra CO_2 fluxes and their role in global climate, which is what the paper ended up being about. But if the story ended up with resolution 2, readers would be dissatisfied. It ended up being about soil microbial communities—a bait and switch that wastes readers' time.

On the other hand, if you frame the opening narrowly and then end up with a wide story (figure 5.1c), you undersell yourself. For example, imagine if the story were about modeling CO_2 fluxes in the tundra (resolution 2), but you opened the paper this way:

"Bacteria living in tundra soils are well acclimated to surviving the cold conditions of the Arctic winter."

This promises a story about the physiology of tundra bacteria, not something that would interest someone focused on the global C cycle. This would turn off a community of potential readers who might have been interested in your conclusions.

Frame your opening to promise the story you will deliver. If you err, though, it's better to err slightly on the wide side. If you oversell in the immediate opening, you can still filter down quickly. Example 5.11 illustrates this; Weintraub targeted the entire community interested in the role of the Arctic as a storehouse of carbon. We felt that his work should interest that community, and I think we were right— the paper has been well cited in journals ranging from microbial ecology to global biogeochemistry. The first paragraph, however, makes it clear the paper is about carbon and nitrogen interactions in tundra soils. Though we tried to convince the biogeochemistry community that they should care about this, if they don't, they can stop reading there. If you frame too narrowly, you lose readers immediately, and once lost, you can't get them back.

5.7. POSITIONING STATEMENTS: PAWN-PUSHES VERSUS QUEEN-LAUNCHES

One student I know says she hates first paragraphs, particularly first sentences, because they usually say little, offering standard platitudes rather than insight. In poorly written papers, that is true. Inexperienced writers often imitate opening lines and come up with platitudes. But in a well-written paper, the sentence may be more—it may be a careful positioning statement that is critical to building the story. How can a sentence be an effective opening in one paper and a throwaway line in another?

Let me answer that with the analogy of a chess game. There are only 20 possible first moves in chess. The most common is to advance the king's pawn two spaces (pawn to king 4; figure 5.2a). Beginners and masters both start games with this move. But it isn't really the same. When a master pushes that pawn forward, it is a carefully thought-out positioning move, the start of a sequence designed to take control of the board and define the structure of the game. Beginners, on the other hand, may push the pawn because they have seen their betters open that way and have a vague understanding that it is a "good" move, but without a sense of what they intend to follow it with.

Though chess is limited to 20 first moves, writing is not. You don't have to push a pawn. If you want to open by launching your queen into the middle of the board, you can (figure 5.2b). You can start a paper with a strong statement that dives in to take control of the story.

For examples of these different approaches, consider first, example 5.3, the Sklar and Dietrich paper on river incision. They used a two-sentence opening. The first was an undramatic pawn push, but it was carefully designed, allowing them to introduce erosion at the end of the sentence—a power position. That was an essential first step to prepare for their next one, introducing stream incision. In contrast, in example 5.8, Azam and Worden launched a queen with their opening about van Leeuwenhoek and Viking.

When you write a straight OCAR story, as is common for specialist journals, you can use a pawn push—an opening that unfolds for a patient audience. If you're writing for *Nature* or the National Institutes of Health, however, you are likely using an ABDCE or LDR structure that start with action, so you had better launch a queen.

William Zinsser argues that the most important sentence in any article is the first one. Yet the student I mentioned thinks most opening sentences are a waste. So who's right? They both are. Most openings are poorly done and unnecessary pawn pushes. That doesn't make Zinsser wrong. To write well, you need to learn how to use the power of the opening. Learn when to use a pawn push and when to launch a queen. Learn to push a pawn like a chess master—as the first step in a strategy to develop your argument and take control of the game. Remember the words of Aristotle, "Well begun is half done."

EXERCISES

5.1. Analyze published papers

Evaluate the openings of the papers you are analyzing. Did they do a good job of identifying the larger issue? What style of opening did they use? Was it a pawn push or a queen launch? Did they dive straight in, targeting a narrow audience, or did they use a two-step approach to engage a wider audience?

Figure 5.2. A pawn push versus a queen launch. A queen launch isn't possible in chess, but it is in writing.

5.2. Write a short article

Evaluate the opening of your short piece and those of your writing group members. Who is the intended audience? Was the opening effective? If not, can you rewrite it to make it so? Is the opening a pawn push or a queen launch? If it was a pawn push, could you write it as a queen launch?

5.3. Revise the following to make them more effective openings. Make the direction clearer and more engaging:

A. The rates of all chemical reactions increase with temperature. This phenomenon grows directly from physical chemistry's transition state theory and the Arrhenius equation. However, respiration in soil doesn't always appear to follow this pattern. Some studies have shown no respiration response to increasing temperature, while a few have even reported a negative response.

B. Chemotherapy is a dominant treatment approach for many types of cancer, and with the development of new targeted-delivery systems has the potential to become even more widespread and efficacious. A common constraint to effective chemotherapy is, however, patient resistance to the treatments. Such resistance is often closely associated with the activity of the enzyme γ-glutamyl transpeptidase (GGT), which acts to increase intracellular concentrations of glutathione and thereby block the apoptotic cascade in tumor cells. Inhibiting GGT before chemotherapy would therefore reduce tumor cell resistance and increase treatment effectiveness.

6

The Funnel: Connecting O and C

Our task, your task . . . is to try to connect the dots.
—Donald Rumsfeld

The opening of a paper identifies a large problem, while the challenge defines a specific question. The main body of the Introduction must connect these elements. It forms the funnel in the hourglass; it narrows the focus and leads readers from the general to the specific, drawing them along the story and framing in the knowledge gap. This is where you build the argument that to make progress on the large problem, you must answer the specific questions.

When you frame the knowledge gap, you provide the background information necessary to understand the story. In an OCAR structure, the background material flows seamlessly from the opening—it is an extension of introducing the problem and the main characters, which is why I don't call it a separate section (hence OCAR, instead of OBCAR). This is in contrast to an ABDCE structure, where after the initial action, you must back up and fill in the background before moving into the development, creating a distinct story element.

Framing the knowledge gap taps into core elements of the SUCCES formula for a sticky story, particularly the U and E elements, unexpectedness and emotion. By defining a knowledge gap, unmasking a hole in the wall of knowledge, you create unexpectedness: I didn't realize that we didn't know that! By closing with a question, you create curiosity: what is the answer? Then you can tell us how you solve the problem and satisfy our curiosity.

If you do this well, you can bridge from very large problems to very narrow questions. For example, you can argue that to understand the global climate system we need to study bacteria in the frozen soils of the arctic tundra during the winter,[1] or that to cure coronary artery disease in adults it is valuable to map the distribution of ISL1+ cells in early fetal hearts.[2] If you do this badly, expect a rejection letter.

6.1. EXAMPLE OF THE FUNNEL AT WORK

Here is the Introduction from an important paper in atmospheric chemistry.[3] This is an extreme example of narrowing the funnel. It opened with a problem at the global scale (global warming), but the research defined the rate constant of a single chemical reaction. The paper had to convince readers that this extraordinarily constrained piece of laboratory research made a contribution to understanding the global climate system, which it did. That required a careful exercise to connect from the global to the molecular.

In example 6.1, I identify important points with numbers in curly brackets (e.g., {1}), and I eliminated references to make it easier to read the text.

Example 6.1
{1} Of all the trace tropospheric species (that is, excluding H_2O and CO_2) methane contributes most to the infrared heating of the atmosphere. {2} Methane is also the most abundant hydrocarbon in the troposphere where it modulates the concentration of the OH free radical and serves as a source of CO. {3} Transport of methane to the stratosphere provides a termination step, via the $Cl + CH_4$ reaction, for the chlorine-catalyzed destruction of ozone. The oxidation of methane in the stratosphere is an important source of water vapour in this region. During the past decade the abundance of methane in the troposphere has been increasing at a rate between 16 and 13 parts per 10^9 volume (p.p.b.v.) per year. {4} The total input and the identities and strengths of the different atmospheric methane sources are not clearly defined. {5} To understand the atmospheric effects of methane, and possibly to regulate it, we need these parameters. {6} At present, the total flux of methane into the atmosphere is estimated from the measured steady-state

1. J. P. Schimel, "The Bugs of Winter: Microbial Control of Soil Biogeochemistry during the Arctic Cold Season," National Science Foundation (2004).

2. Bu et al., "Human ISL1 Hear Progenitors Generate Diverse Multipotent Cardiovascular Cell Lineages," Nature 460 (2009): 113–17.

3. G. L. Vaghjiani and A. R. Ravishankara, "New Measurement of the Rate Coefficient for the Reaction of OH with Methane," Nature 350 (1991): 406–9.

abundance and the known removal rate of methane. It has been generally accepted that the only process by which methane is chemically degraded in the troposphere is the reaction with OH. {7} Therefore, the rate coefficient, k_1, for the reaction

$$OH + CH_4 \rightarrow CH_3 + H_2O \qquad\qquad (1)$$

is important in estimating the total flux of methane. The other loss processes, which are expected to be minor pathways, are surface deposition and reaction with Cl atoms in the lower stratosphere and upper troposphere.

{8} A close examination of the available data shows that only in three investigations was k_1 measured below 298 K, the temperature region most important to the atmosphere. Only Davis *et al.* measured k_1 down to 240 K. Reaction 1 is slow. Therefore, at low temperatures, the presence of reactive impurities and occurrence of secondary reactions in laboratory systems can result in an overestimate of k_1. {9} We studied reaction 1 using an experimental method in which secondary chemistry could be minimized and the systematic errors reduced.

{1} This is the opening, which frames a story about atmospheric methane (CH_4) and the greenhouse effect. This reaches for a wide audience—it includes anyone interested in global warming, which definitely includes *Nature* editors and readers.

{2} The second sentence introduces the other critical character in this story—OH (hydroxyl radical). By pointing out the CH_4 "modulates" OH, without discussing OH, the authors take for granted that you know why OH is important (a necessary weakness in a paper this short).

{3} This section adds information about why CH_4 is important in the global system. However, because the main story line goes from CH_4 to OH, this material may seem out of place—it goes back to the opening about why CH_4 is important. But the authors presumably felt it important to introduce OH radical as a character early on. Thus, this structure creates an ABDCE story line. The first two sentences formed the A part, and now this backs up to fill in the background (B).

{4} This is a critical statement in laying the base of the knowledge gap: "total input and . . . methane sources are not clearly defined." This paper is going to more clearly define them.

{5} This helps establish the importance of the research—we need to fill the knowledge gap to better manage sources and sinks and mitigate the role of CH_4 in causing global warming.

{6} Here is another critical point in establishing the knowledge gap. The sources of CH_4 are hard to measure, but the major CH_4 sink is reaction with OH radicals, so we can estimate CH_4 fluxes into the atmosphere as being equal to the losses via reaction with OH radicals. This brings the OH radical—a central character—back into play.

{7} At this point, the authors have narrowed all the way down to the molecular scale and the importance of knowing the rate constant for this reaction: to understand the total flux of CH_4 to the atmosphere, we need to know the rate of its reaction with OH, and that means we need to know the rate constant for that reaction. This is essential to understanding the overall role of CH_4 in global warming.

{8} Here, they finish defining the knowledge gap. Having established that we need to know the rate constant k_1, the authors tell us that only three studies have tried to measure k_1 at realistic temperatures, and only one has done so at a temperature that by implication, is in the right range for atmospheric reactions. We need better measurements of k_1 at realistic temperatures to understand atmospheric CH_4 dynamics. These authors quickly scaled down from the global to the micro scale and did so in a way that, at each step, identified what we needed to know. They only tell us what we know to define the limits of that knowledge, rather than for its own sake.

{9} This is the specific statement of the challenge, and unfortunately, I think it's a dud. After clearly framing the knowledge gap and its importance, the authors stated their challenge by saying "We studied Reaction 1 . . ." It's obvious that the question is "What is the value of k_1?" But it would have defined the knowledge gap more concretely to say, "We measured the value of k_1 at temperatures down to 230 K using an experimental . . ." This weak challenge highlights an important point: you don't need to be perfect to be successful. This was an important paper.

This was a *Nature* paper and so quite condensed—it had to narrow quickly with broad strokes. In papers for specialist journals, you have more space to develop the Introduction and can do the narrowing more gently and thoroughly. You probably won't have as imposing a task either, narrowing all the way from global to molecular scales. However, the stepwise narrowing process will be the same—make sure that you aren't telling us everything you know about a topic but developing the logical connections between each step to frame the knowledge gap.

6.2. BAD INTRODUCTIONS: FAILING TO DEFINE THE PROBLEM

A good Introduction defines a problem and narrows to an interesting question. A weak or poor Introduction, in contrast, either fails to define the problem or tries to sell a solution before defining the problem, and so fails on curiosity.

6.2.1. Failing to Identify the Problem

Many papers are unclear in defining the problem. They introduce it, tell us that "little is known about this topic," give us some information about it, and close the Introduction by saying "our objectives were to carry out the following tasks." Such works are common in an editor's "New Submissions" folder but are much less frequent in her "Accepted" folder.

The problem with this style of Introduction is that it does a poor job of defining the problem or the value of the solution. It's not very convincing to say "little is known about X" for scientific, logical, and literary reasons.

Scientifically, it is unconvincing because it's probably false. Very few of us have written a paper on a topic that hasn't had tens or hundreds of studies already published on it. Invariably, we know a *lot* about the topic at hand. There are important questions remaining, which is why we did the work, but those are bounded and defined by a large body of knowledge. So when someone says, "little is known about X," we often feel that the author either doesn't know the literature or is overstating the case.

Logically, it's unconvincing because after saying "little is known," the authors describe a lot that *is* known. Even if the short list of facts is everything known on the subject, it comes across as a data dump that contradicts the argument. We don't see the "little."

Finally, linguistically, it's not convincing because it's not concrete. "Little is known" is fuzzy—how little is little? If you tell us the six things that are known, is that still a "little?" Because the language is fuzzy, the argument is unconvincing. To make it convincing, it needs to be concrete—what specifically do we not know?

You must explicitly define the problem, as illustrated in example 6.1. They didn't say that "little is known about CH_4 sources." That would have been inaccurate; we knew a lot about CH_4 sources. Rather, they said "sources are not clearly defined," which is tighter language that implies something closer to "while the broad patterns are known, important details are not," a true description of the state of knowledge at the time and enough to get the paper into *Nature*. A concrete statement that defines a small knowledge gap will do better than a fuzzy one that fails to define one.

6.2.2. Offering a Solution before Defining a Problem

Sometimes authors offer a solution before defining the problem. As you are working on a paper, you live with the topic so closely for so long that it is easy to assume that the question is obvious. It can become hard to see that you haven't posed it clearly. As a result, authors sometimes end up taking the problem for granted and focus on their solution. This creates what I call the "bizzwidget problem" as illustrated by a scenario with a door-to-door salesman: "Hi, ma'am, I'm selling the new Buzco Bizzwidget. The Bizzwidget is the most amazing tool you've ever seen—why, I don't know how you've ever lived without it! So here, let me show you some of the wonderful things it does."

Ma'am is already trying to get rid of the salesman—without finding out that the Bizzwidget really is an amazing tool that she might want to buy. He's trying to sell her a solution, but she doesn't know she has a problem to solve. This strategy works with customers who are inordinately patient, but mostly with people in the "Bizzwidget community" who already know how wonderful the product is.

For the rest of us, we're with the hapless "customer," and the salesman is on the street staring at a closed door. If you are trying to sell us a bizzwidget solution, first convince us we have a problem: "Hi, ma'am—have you experienced problem X? You have? Do you have a solution? You don't? Well I do—let me show it to you; we call it the Buzco Bizzwidget."

Now the Bizzwidget isn't mumbo jumbo. Importantly, this approach engages anyone who has ever experienced problem X, not just the few who have heard of the Bizzwidget. It does this by opening with a concern many people share (defining the audience in the opening), and then *showing* us why we need a Bizzwidget (the body of the Introduction), before introducing the specific product (the challenge). This approach engages our curiosity—do you have a solution? How does it work? It is also concrete—it identifies a real problem and its solution.

If you don't recognize the bizzwidget problem in science writing, consider the following example.

Example 6.2:
Addressing complex interactions among chemistry, physics, and biology in climate systems requires an interdisciplinary approach. We propose to address this challenge by using Complex Systems Modeling Theory (CSMT). CSMT has been used in chemical systems to model molecular reaction mechanisms and in cell biology to model physiological pathways. It has been used . . .

This uses the bizzwidget approach; it assumes that we know the problem that CSMT is the solution to and so doesn't define it. It doesn't describe the complex interactions, how other approaches have struggled with them, what the CSMT approach is, or why it is better than other approaches. We may find all that out later in the paper—that CSMT is a solution to a problem we care about—but unless our neighbor already has told us about CSMT, we've probably closed the door on this one. To sell us a solution, first sell us a problem.

6.3. INTRODUCTION VERSUS LITERATURE REVIEW

The need to narrow the focus and lead the reader to your specific questions means that an effective Introduction cannot be merely a literature review that synopsizes what we know about a topic. Instead, because you must convince us of the importance of the problem, you must show us what we *don't* know and why it is important.

The difference between a literature review and an Introduction can be subtle, because they both use the existing literature to discuss the state of knowledge. The distinction between them is that a literature review builds a solid wall—describing knowledge—whereas an Introduction focuses on the hole in that wall—describing ignorance. They tell different stories and move the story in different ways. They also use the existing literature differently; an Introduction focuses on the publications that define the edges, rather than the core of knowledge.

How do tell when you are writing a literature review rather than an Introduction? See whether you are focusing on telling us what we know or what we don't. When you describe something we know, do you use it to identify the boundaries of that knowledge? If so, you're writing an Introduction; if not, you're probably creating a literature review.

One clear flag for when you're doing a literature review is when your citations are at the beginning of sentences. Do you write: "Smith (2003) found X" or do you write: "X occurs (Smith 2003)"? The former tells a story about Smith and what she did; the latter, about nature and how it works. If you write the former, you are probably doing a data dump, collecting the information that seems relevant and writing it down, without synthesizing it and integrating it into a story or framing a knowledge gap. The important information is almost never that Smith found it; rather, it is almost always *what* she found. So why make Smith the subject of the sentence? Whenever you see that you've written a "Smith found . . ." sentence, ask whether the researcher, rather than the research, is what you want to tell us about. If not, rewrite it to focus on the findings. Doing this will help you tighten up the arguments and sharpen the knowledge gap.

There are cases in which you might want to highlight the researcher. The first is when you are discussing an ongoing debate: "Although Smith (2003) reported X, Jones (2005) found Y." This highlights that there is no agreed-on truth but a collection of individual opinions. If there is an accepted dogma that one researcher is challenging, however, you would write something like: "While most reports suggest X (e.g., Smith, 2003, Xu 2004), Jones (2005) found the opposite, arguing . . ."

When there are two camps with multiple papers supporting each side, it is probably best to condense and synthesize it all to "There is still uncertainty about the nature of X, with some reports suggesting it is Y (Smith 2003, Xu 2004) and others suggesting it is Z (Arif 2005, Masukawa 2006)."

The "Smith found X" approach to discussing the literature is common; I do it all the time in my early drafts. But it frequently signals that we haven't fully synthesized the information and figured out why we're presenting it. It is a flag that we're still in the data-dump stage and need at least one more major revision.

Most OCAR papers use a simple O → C flow in the Introduction, with a smooth funnel from the opening to the challenge to define the knowledge gap. In contrast, most proposals use a structure more akin to ABDCE. They have an opening section that makes the overall case for the work and briefly sketches in the knowledge gap. Then the background sharpens and fills in that sketch to justify the proposal's specific challenge. But that background is not a literature review—it must still be an introduction. The background is *never* a place for a data dump where you tell us everything about the field. If a piece of information does not have a specific and concrete role in moving the story forward, it does not need to be included.

The vital elements of an Introduction are the opening and the challenge. Those are the "dots" that you must connect by filling in the background and forming the funnel. That material has only one purpose: to show a reader why answering your questions is essential to making progress on the overall problem. By the time readers reach the challenge, they should feel that your questions are the obvious ones, even if they had never thought about them before.

EXERCISES

6.1. Analyze published papers

Go back to the papers you've been analyzing. Look at their Introductions and determine whether they frame the knowledge gap effectively. Does the Introduction have a clean funnel that flows from the opening problem to the specific questions?

6.2. Write a short article

Look at your short article and those of your group. Evaluate the funnel part of the Introduction—does it frame the knowledge gap? If not, revise it so that it does.

7

The Challenge

Scientia (Latin): knowledge.

In the challenge, you describe the specific knowledge you hope to gain. This starts with the question that drove you to do the research. You did the work to discover the answer. From the question, we sometimes formulate a hypothesis and we usually state specific objectives, which describe the information we will present. Some authors only pose the question, whereas others do all three, offering a question, framing it into a hypothesis, and then describing specific research goals. Each approach has its place, but the question is the core of it all. If you don't have a question, you are not doing good science. If readers can't tell what it is, you are not writing good science.

7.1. QUESTIONS VERSUS HYPOTHESES

There are people who argue that without a hypothesis, it isn't science. That view grows from a strict focus on Popper's argument that a theory is only scientific if it can be falsified.[1] But Vaghjiani and Ravishankara (example 6.1) had no hypothesis

---

1. K. Popper, *The Logic of Scientific Discovery* (1934; Routledge Classics, 1959).

about the rate constant for the reaction between methane and hydroxyl radicals; they did have a question, "what is the rate constant?" Was their work not science?

Different fields of science have different traditions about questions versus hypotheses. One community I work in submits proposals to the National Science Foundation Biology Directorate and demands hypotheses. Another submits to Geosciences and is slightly baffled by biologists' obsession with them. Framing a hypothesis can be a powerful tool for organizing your thoughts and structuring your research, but a hypothesis merely takes your question and makes into a falsifiable prediction. The question, defining the knowledge gap, is still the key.

We often use hypotheses to test whether a relationship exists—to develop a theory. We don't necessarily frame one to evaluate the nature of a relationship. An evolutionary biologist might ask if flower color controls whether pollinators visit a particular plant—they might frame the hypothesis: "hummingbirds prefer red flowers." That is testable and falsifiable. But Vaghjiani and Ravishankara knew that methane reacts with hydroxyl radicals and that the reaction has a rate constant, they just didn't know what it is. What would they have hypothesized?

Interestingly, even biologists who routinely frame hypotheses for proposals might still ask a question in the challenge of a paper. I can't give any global advice as to whether to pose a question or to transform it into a formal hypothesis; you need to know the culture of your field. Just remember that the question comes first and must be clear.

7.2. QUESTIONS VERSUS OBJECTIVES

Despite the importance of the question, many authors define their challenge by stating "Our objectives were" rather than by saying "Our question was." That is, they focus on the information they will collect, rather than the knowledge they hope to gain. They assume that the question is obvious from all they have said in the introduction and they don't need to state it explicitly. They are almost always wrong.

Focusing on objectives instead of questions is weak science and weak storytelling. If you leave the question unstated and implicit, and jump straight to specific data-collection goals, the reader has to figure out what your question was and whether you even had one. You leave it to them to figure out how the work will advance knowledge. That violates principle 1—it is the author's job to make the reader's job easy.

Focusing on objectives also doesn't engage SUCCES. It doesn't create unexpectedness or curiosity—at least not the curiosity you want. A reader will wonder about your objectives. Why did you do this work? What was the purpose? What was the underlying question these tasks address? Is it possible that you were just aping experiments published by others without a real question of your own? Those are not questions about your science but about you and your motivation, and there is an underlying criticism embedded—why are you wasting my time with this?

You have a question that drove your work. Make it clear. Then you can tell us how you answer it.

7.3. WHAT COMES AFTER STATING THE QUESTION?

After posing the question, a good challenge briefly lays out the research approach. This is where you tell us about specific objectives and the information you will generate. If you tested whether a pollutant is carcinogenic, did you use mechanistic toxicology or epidemiology? If you were identifying bacteria in nature, did you grow them in culture or sequence DNA extracted from environmental samples? If you are measuring the rate constant for a reaction, did you do it in gas phase or aqueous solution? Here, stating objectives can be useful in providing a map that helps readers assess the rest of the paper. Were your methods appropriate for the question? Did you learn what you hoped to? Where are the remaining gaps?

Some papers also provide a brief overview of the Conclusions, usually starting with language like "In this paper we show . . ." After a full Introduction, the authors telegraph their Conclusions. This telegraphed OCAR works well for an impatient audience but one that still wants to see how arguments develop. It is more frontloaded than normal OCAR, but not as much as ABDCE or LDR. This approach is common in the biomedical literature, but is not unknown in other fields. I suspect it evolved to help readers screen papers quickly—knowing the proposed conclusions makes them easier to evaluate as you read.

7.4. GOOD CHALLENGES

In papers for specialist journals, a good challenge almost always condenses conceptually to "to learn X, we did Y." That is, they present the question and lay out an approach to answering it, as illustrated in example 7.1. This paper explored how living in a complex environment may enhance brain development.

Example 7.1
Our goal in this study was twofold. First, we tested whether an animal's physical environment would affect hippocampal attributes. Specifically, we tested whether food-caching mountain chickadees (*Poecile gambeli*) housed in captivity differed in hippocampal volume, hippocampal neuron number and neuronal density as compared with fully developed wild-caught conspecifics. We predicted that captivity, with reduced environmental complexity and restricted memory-based experiences (compared with memory-based experiences afforded in the natural environment), would reduce hippocampal volume, neuron number and, potentially, neuron density.[2]

2. L. D. LaDage, T. C. Roth, R. A. Fox, and V. V. Pravosudov, "Effects of Captivity and Memory-Based Experiences on the Hippocampus in Mountain Chickadees," *Behavioral Neuroscience* 123 (2009): 284–91.

Why is this a good challenge? It is long and technical, without linguistic flourish. The authors, however, do several things well. First, they remind us of the overall issue—fundamentally a question about controls on brain development, a topic of wide interest and import, even though the specific question is narrow. In fact, the specific question about captive birds and their hippocampuses, by itself, might seem more a candidate for a Golden Fleece award for pointless Ivory Tower science than for the Faculty of 1000 website of interesting and important papers, which is where I found it. The authors did an excellent job of connecting their specific question to the larger problem. They go further, though— even after posing a tight question, they frame a hypothesis that defines their measurements and the data that would falsify or support it. As you read the rest of the paper, you know what the authors think and what they did; you can easily follow along as they assess their results and develop their Conclusions. Nicely done.

Example 7.2 is from a paper looking at the mechanisms of light perception and visual responses in marine invertebrates. What chemicals create the signal? Some work suggests that diacylglycerol (DAG) has a role in signaling. But the authors argue that DAG can't completely explain observed responses and so propose an alternative signal molecule: phosphatidylinositol bisphosphate (PIP_2), which breaks down to DAG.

Example 7.2
Despite the tantalizing evidence for DAG and/or its downstream products in visual transduction and the synergistic role of calcium, in no instance has application of such chemical stimuli fully reproduced the remarkable size and speed of the photocurrent. This may imply that yet another signal may be missing from the proposed schemes. In other systems PIP_2 has been shown to possess signaling functions of its own, independent from those of its hydrolysis products. . . . These observations prompted the conjecture that in microvillar photoreceptors PIP_2 may help keep the channels closed and its hydrolysis could promote their opening. In the present report, we examined the consequences of manipulating PIP_2 on membrane currents and light responsiveness in is qolated photoreceptors from *Pecten* and *Lima*.[3]

The authors clearly lay out the problem—DAG can't explain existing observations. They hypothesize a new mechanism that involves PIP_2 and briefly describe the experiments they did to test this hypothesis— "to learn X, we did Y." Even though this was in a specialist journal, the authors use strong words to engage curiosity and attention, notably *tantalizing* and *remarkable*.

As an example of a telegraphed OCAR structure, consider example 7.3. This paper explores how the transcriptional machinery at tRNA genes may interfere

---

3. M. Del Pilar Gomez and E. Nasi, "A Direct Signaling Role for Phosphatidylinositol 4,5-Bisphosphate (PIP2) in the Visual Excitation Process of Microvillar Receptors," *Journal of Biological Chemistry* 280 (2005): 16784–89.

with DNA replication in a way that can promote chromosome breakage. They examined how certain proteins regulate tRNA gene transcription.

Example 7.3:
Although Rrm3 and Tof1 might collaborate to set the rate of fork progression through tRNA genes, there is no evidence that this rate is subject to physiological regulation by mechanisms that determine the balance of activity between Rrm3 and Tof1. Furthermore, tRNA gene regulation is not known to be tied to fluctuations in the rate of DNA replication other than by mechanisms that generally tune the proliferation rate to nutrient availability and overall cellular fitness for growth and division. However, there is evidence that the appearance of double strand breaks (DSBs) in DNA can trigger repression of the tRNA genes. Specifically, tRNA gene transcription is actively repressed in cells treated with UV light or the alkylating agent methane methylsulfonate (MMS). The secondary DNA lesions generated in UV-irradiated and MMS-treated cells include DSBs. This fact, as well as the observation that tRNA gene repression requires a protein kinase (CK2) previously implicated in adaptation to chromosome breakage, led us to hypothesize that the canonical DNA-damage response (DDR) checkpoint pathway controls tRNA gene transcription. The aim of the present study was to test this hypothesis. In pursuing this aim, we discovered an unexpected system of regulation in which checkpoint proteins specialized for signaling replication stress repress tRNA gene transcription during normal proliferation. These proteins also convey repressive signals to tRNA genes in cells exposed to genotoxins that cause replication interference. These data provide the first evidence that the fork-pausing activity of tRNA genes is regulated by the checkpoint system that has evolved to control replication.[4]

The authors synthesize existing knowledge to pose a clear hypothesis; they could have stopped by saying that their aim was to test that hypothesis. Had they done so, this still would have been a fine challenge. But then they highlight that in testing the hypothesis they discovered a new regulation system, in which the mechanisms that control the rate of DNA replication also regulate tRNA transcription. This both prepared you for the story to come and raised the curiosity factor to intensify the challenge.

These examples all came from papers in specialist journals written using OCAR story structure and IMRaD sections; they all used the "to learn X, we did Y" form. When a paper uses a different structure, the challenge is often condensed to focus more intensely on the question. Example 7.4, from the field of physical chemistry, illustrates this in an LDR-structured paper; it is a report in *Science* that doesn't use subheads to break up the sections. The authors ended the paper's lead with a

4. V. C. Nguyen, B. W. Clelland, D. J. Hockman, S. L. Kujat-Choy, H. E. Mewhort, and M. C. Schultz, "Replication Stress Checkpoint Signaling Controls tRNA Gene Transcription," *Nature Structural and Molecular Biology* 17 (2010): 976–81.

strongly worded question to ensure that you didn't miss the challenge and to effectively engage the broad audience of this journal.

Example 7.4:
However, three decades of work in the gas phase have explored how the specifics of the forces between atoms involved in isolated chemical reactions determine the final energy partitioning as the reaction moves from the transition state. Is knowledge of these specifics completely immaterial to reaction dynamics in solution?[5]

These authors define the knowledge gap and ask an interesting question that confronts the reader—is the three decades of work on gases "completely immaterial" to solution-phase reactions? This taps into SUCCES: it draws on simple by asking a clean and straightforward question, and it draws on unexpected and emotion by using the highly charged phrase "completely immaterial" to challenge decades of high-quality science.

7.5. BAD CHALLENGES

If the challenge is unclear, readers will be left adrift. If they don't know where the paper is going, how will they know whether they got there? A challenge is ineffective if it doesn't concretely state the question or hypothesis or if gives the reader the wrong impression as to what it is.

The most common type of unclear challenge is where authors focus on the information, rather than the knowledge they are trying to acquire; they leave off the "to learn X . . ." and just say "we did Y." They focus on the objectives, rather than the question. I think many authors fall into this trap because they know the material so well that the question seems obvious. No reader, however, knows as much about your work as you do, and your thinking is rarely completely apparent. The challenge is too important to leave it to assumption, hope, or chance. You must make the question clear. If you fail to do this, your papers will lack power, and your proposals will likely lack funding.

So let us evaluate some weak challenges and discuss how to improve them. The first example explores why the immune system sometimes breaks down.

Example 7.5
Some T-cells may be anergic—that is, unable to proliferate after being restimulated with an antigen. Some anergic T-cells are unable to link to the T-cell–antigen presenting cell (APC) interface. Here we examined the structural

5. A. C. Moskun, A. E. Jailaubekov, S. E. Bradforth, G. H. Tao, and R. M. Stratt, "Rotational Coherence and a Sudden Breakdown in Linear Response Seen in Room-Temperature Liquids," *Science* 311 (2008): 1907–11.

WRITING SCIENCE

characteristics of anergic mouse T-cells and we tested their functional response to being rechallenged with antigen-loaded APCs.

Here the authors tell us which data they will collect, but they don't specify the knowledge gap. What is the question? Presumably these researchers are trying to figure out what makes a T-cell anergic, and from that to understand why immune systems break down, with the ultimate goal of finding ways to prevent the breakdown. But that thought process is opaque—the implicit question is too deeply buried. This would have been much stronger if they had clarified the "to do X" part of the challenge, perhaps like this:

> "To determine what causes mouse T-cells to be anergic, we evaluated the structural characteristics of T-cells and how they responded to being rechallenged with antigen-loaded APCs."

That simple addition would have clarified the question and made it stronger.

Example 7.6 is from environmental science and is about how herbivores structure plant communities.

Example. 7.6
We evaluated the possibility that hares influence the structure of shrublands by acting as keystone herbivores that maintain gaps between the shrubs and so influence the competitive interactions of plants recruiting into those gaps.

In this one, the question itself is unclear. What do hares do that influences shrubland structure and competitive interactions? What does the investigator hope to learn?

There are several implied hypotheses within this challenge: (a) hares control plant community structure, (b) they maintain gaps between shrubs, and (c) they influence competitive interactions within the gaps. Those hypotheses should be explicit and concrete. This should also describe the experiment that will test those hypotheses. Consider this as an alternative:

> "We hypothesized that hares control the structure of shrublands by foraging on shrub seedlings in the gaps between mature plants. If true, hares act as keystone herbivores by maintaining these gaps, in which grasses can outcompete shrub seedlings. We tested this hypothesis by following hare movement to determine where they eat and by analyzing their feces to determine what they eat."

This states the hypothesis and suggests its larger implications—hares are keystone species that have a disproportionate impact on ecosystem structure and function. It also briefly identified the experimental approach used to test the hypothesis.

Example 7.7 is also from environmental science, specifically grassland ecology and global change biology. The authors pose two goals, both of which are clear. The problem is their sequence.

Example 7.7
The study had two goals. First, we aimed to constrain our estimates of grass-land plant production by comparing measurements based on two techniques: maximum biomass at the end of the season and periodic measurements of photosynthesis. Second, we examined the response of grass growth to a combination of elevated CO_2 and increased temperatures, conditions that are expected to occur with climate warming.

Why is this challenge weak? The authors presented the goals in an order that is more chronological than intellectual. First they validated their methods, and then they assessed what the data meant (i.e. We did X to learn Y). But we expect the most important objective to come first, defining a study's overall thrust. Following objectives should elaborate and refine that main goal. Starting with the data-collection goal gives readers the impression that the work is primarily comparing approaches for measuring plant growth—narrowly useful, but not broadly engaging.

The interesting question is posed in the second objective: how will plants respond to climate change? This is the *To learn Y* part of the challenge. Because it comes second, it seems subordinate to the methods comparison. The language reinforces that hierarchy; the first goal is stated using strong verbs: the authors will "constrain" and "compare" measurements. The second, in contrast, uses weak language: they will "examine" a response. To fix this problem, we need to switch the order of the objectives and highlight the core question:

"The primary goal of this study was to evaluate how grass growth responds to a combination of elevated CO_2 and increased temperatures, conditions that are expected to occur with climate warming. To validate our plant growth measures, we used two approaches to estimate plant production: maximum biomass at the end of the season and periodic measurements of photosynthesis."

Now this seems like a more interesting paper—the real question is clear and is unmistakably about a topic that is broadly interesting and relevant.

A good challenge must define not only the data you collect but the knowledge you hope to gain. If you read something and can't find a clear statement of the question or hypothesis, or if that question itself is unclear, the challenge will be weak and will weaken the entire story. With a proposal, that weakening is likely to be fatal. Remember that the critical part of the challenge is not "we did Y" but "to learn X."

EXERCISES

7.1. Analyze published papers

Evaluate the challenge of each paper. Does it clearly frame the question? Could you write it better? If so, how?

7.2. Write a short article

In your short article, did you clearly pose the question? If not, rewrite the piece to do so.

8

Action

You are not just presenting your results, you are telling a story.

In OCAR, action makes up the main body of the story and includes everything between the challenge and the resolution. In a paper, this includes the Materials and Methods, the Results, and most of the Discussion. In a proposal, it is the description of what you intend to do. Because so much goes in these sections, you could write an entire book on it (and many have). Particularly when it comes to how to present data (tables, figures, etc.), there is a wealth of information. Because this is a book about *writing*, I focus on how to integrate these sections into the overall story and how to use story structure to present them most effectively.

In writing the action, the critical message is to remember the last S in SUCCES—*story*. You are not just presenting your results, you are telling a story. You are, of course, free to write papers that simply present experiments and data; but journals are equally free to reject them. It's not that readers aren't interested in your techniques and results—we are. We want to know what you did and what you found. That is the concrete core of the science, and we must be able to evaluate it to assess the validity of your conclusions. But without embedding the action within the larger story, the paper easily becomes aimless, incoherent, and dull. What is the point of all that work? What do these results mean? Do they answer your question? Do they support your hypotheses and conclusions?

By integrating the action into the story, you give it structure and direction. You help the reader work through the results to figure out what they mean and how they fit together. Let the story guide you in describing your methods, in choosing which results to present, and in how to present them. As you write each section, think about it as a mini-story that has its own OCAR elements. Consider why you are telling us any particular piece of information—how it contributes to the larger story—and frame the presentation to highlight why you did something as well as what you did.

The action sections of a paper can be separated into two distinct parts: describing what you did (Materials and Methods) and what came of it (Results and Discussion).

8.1. METHODS

A principle of science is that other researchers should be able to repeat a piece of work, which means a paper must offer a detailed explanation of procedures. Few of us, however, are likely to repeat a piece of work; more often we evaluate the methods to assess the credibility of the data and conclusions. What we need is an overview of the information you were trying to gain and the approach you used to gain it. Most readers don't want the detail in their first reading and are impatient—so much so that many journals now put the Methods and Materials section at the end of the paper. To serve the needs of all possible readers, the best way to describe a method is use a lead/development (LD) structure, providing an initial overview for all and then the details for those who need them.

For example, consider the following three ways to describe a method. I've seen each of these approaches used many times in papers, but which is best?

Example 8.1
Enzyme Inactivation following 3-HPAA Metabolism
    Enzyme inactivation associated with 3-HPAA metabolism was measured by the method of Turman et al. (2008).

Example 8.2
Enzyme Inactivation following 3-HPAA Metabolism
    PGHS-1 or PGHS-2 was incubated with 25 µM 3-HPAA. When oxygen uptake was complete, arachidonic acid (25 µM) was added, and the maximal rate was determined as described above and normalized to the DMSO control. The concentration dependence of PGHS-2 inactivation was analyzed in a similar manner with varying concentrations of 3-HPAA (from 10 nM to 25 µM).

Example 8.3
Enzyme Inactivation following 3-HPAA Metabolism

To characterize the extent of enzyme inactivation associated with 3-HPAA metabolism, PGHS-1 or PGHS-2 was incubated with 25 µM 3-HPAA. When oxygen uptake was complete, arachidonic acid (25 µM) was added, and the maximal rate was determined as described above and normalized to the DMSO control. The concentration dependence of PGHS-2 inactivation was analyzed in a similar manner with varying concentrations of 3-HPAA (from 10 nM to 25 µM).[1]

Example 8.1 is truly miserable—it doesn't say anything about how the measurement was done. Because the specific method can affect how we interpret results, we could not evaluate this work without tracking down the original paper. Talk about making the reader's job hard! Every paper is supposed stand on its own, and this one would not.

Example 8.2 is imperfect—it goes straight into detail without an opening to guide us in what those details are about. The header provides some structure, but we don't read subheads as part of the sentence. If you have a page limit, you can get away with this, otherwise, try another approach.

Example 8.3 is the best and the one Turman et al. used—it adds a few extra words to provide a brief overview of the goal before describing the approach. The details are there, but the overview is enough to follow the paper. All readers can get the information they need.

The other way to make methods easy to follow is to tap into established schemas. Many methods and techniques are well known in their field, and some have specific names. Use them when available. For example:

We measured protein by the Lowry approach.
We measured microbial biomass by the chloroform slurry approach as
     described by Fierer and Schimel (2003).
We amplified DNA by hot-start PCR.

These start by describing what the authors did (measure protein, biomass, etc.) and then tell us how by naming a well-known approach. For many readers, that is enough. Tapping into the schema of the Lowry protein assay or PCR (polymerase chain reaction) makes the description seem simpler and more concrete. From that base, the author provides the experimental details, but may only need to highlight differences from the standard approach: "We measured protein by a modification of the Lowry protein assay, in which sodium citrate, instead of sodium tartrate, is used in reagent A."

Through such approaches, using LD structure and tapping into established schemas, you can make your methods easier for both novices and experts, allowing them to get the information they need at whichever level they choose.

1. M. V. Turman, P. J. Kingsley, and L. J. Marnett, "Characterization of an AM404 Analogue, N-(3-Hydroxyphenyl)arachidonoylamide, as a Substrate and Inactivator of Prostaglandin Endoperoxide Synthase," *Biochemistry* 48 (2009): 12233–41.

8.2. RESULTS AND DISCUSSION

8.2.1. To Separate, or Not to Separate: That Is the Question

The Introduction and Methods comprise the first half of the paper, where you justify and explain what you did. The second half is where you describe the outcome: your findings and interpretations. You have flexibility in structuring this part of a paper to best present your contributions. Many papers separate Results from Discussion, but others combine them in a variety of ways. However you choose to organize this material, that choice should be grounded in two core principles of writing and science:

1. Make the reader's job easy (our principle no. 1): present results and interpretations in a way that best develops the story.
2. Readers must be able to distinguish what you found from what you think.

I take that second principle further and argue that there are three types of material in a paper:

1. Data: Your actual results.
2. Inference: These are the clear and robust interpretations of the data that almost any practitioner in the field would draw; these are sometimes so obvious that we treat them as data themselves.
3. Interpretation: Your thoughts, hypotheses, and speculation about what the results may mean for the larger problem you identified.

Somewhere in a paper you should address each of these items. Optimizing the mix, however, is a balancing act between the author's taste and confidence and those of the reviewers and editors. Deciding how to structure the presentation is equally a balancing act.

Fields vary in how they deal with results versus interpretation/discussion, depending on the length of papers (short papers tend to integrate) and on the nature of data presented. In fields such as biology and environmental science, where the core information reported is straight data (e.g., chemical concentrations, DNA sequences, etc.), the norm leans toward separate Results and Discussion sections. In more theoretical fields, such as physics, raw data are often processed through theory and models to convert them into the information reported; there is less separation between result and interpretation, so papers tend toward a more integrated presentation.

Using a combined Results and Discussion is also common in observational/ analytical disciplines such as geophysics, where scholars collect information in the field, analyze samples for chemical or isotopic composition, and then use these data to try to reconstruct the history and dynamics of a region. Because each type of information adds an extra dimension to the discussion, it makes more sense to describe a data set and its meaning before going on to discuss the next line of evidence. In modeling papers, it also frequently makes sense to break the presentation into modules, rather than separate all the results from any discussion of them.

Even in experimental fields, there are times when an integrated approach is most effective. For example, when experiments are sequential, with each based on the results of the previous, it can be impossible to intelligibly present all the data before discussing it. Such work usually calls for a more chronological narrative organized by experiment, with each section including both Results and Discussion.

Another time to use a combined approach is when there is a strong integration of data and inference. Consider example 8.4, a paper evaluating a transmembrane protein in *Mycobacterium tuberculosis*.[2] The authors divided the combined Results and Discussion into four subsections.

Example 8.4

Sections within a Combined Results and Discussion

The B and C Domains of *M. tuberculosis* Rv0899 Form Two Independently Structured Modules

Three-Dimensional Structure of the B Domain of Rv0899

Implications for the Biological Function of Rv0899

Implications for the Organization of Rv0899 in the Mycobacterial Membranes

The authors could have split these, putting the first two sections into Results and the latter two into Discussion, but they didn't because the first sections are not just data. They contain analyses that relate to the proteins' specific biochemical functions, which were necessary to discuss but did not develop the larger story about *Mycobacterium*'s pathogenicity.

This is illustrated in the following paragraph from the first section, which weaves together data and inference.

Example 8.5

Paragraph Integrating Results and Inference

Notably, the spectra of Rv0899-B (Figure 3B, red) and Rv0899-C (Figure 3B, blue) form perfect complementary subsets of the spectrum from Rv0899-BC (Figure 3A), spanning both domains, with the exception of some peaks from residues in the BC connecting region. This demonstrates that the B and C domains constitute independently folded modules, as suggested by sequence homology. The resonance line widths measured in the three spectra are very similar, further indicating that all three polypeptides exist as

2. P. Teriete, Y. Yao, A. Kolodzik, J. Yu, H. Song, M. Niederweis, and F. M. Marassi, "*Mycobacterium tuberculosis* Rv0899 Adopts a Mixed R/β-Structure and Does Not Form a Transmembrane β-Barrel," *Biochemistry* 49 (2010): 2768–77.

monomeric species in solution. Since the line widths are not appreciably larger in the spectra of the BC polypeptide, the B and C domains may be significantly dynamically decoupled.

Because the first two sections integrated data and inference, calling them Results would have mislabeled the material. Pulling all the analysis out, however, would have disrupted the paper's overall flow. Instead, the authors used a partially integrated structure that flows from more Result-like to more Discussion-like material. Within this hybrid structure, they distinguished results from interpretation by presenting the data first and by using words like *demonstrates* and *indicating* to identify inferences. This was nicely done.

Imagine, however, if they had written this as follows.

Example 8.6
The B and C domains constitute independently folded modules, as indicated by the fact that their spectra form perfect complementary subsets of the spectrum from Rv0899-BC.

I would characterize this sentence as pure Discussion. Though it refers to the data, it does so to support an argument and the data come after the argument. Whenever the data come first, it feels like you are drawing your interpretations from them, and so are obeying Lamott's dictum of "listen to your characters." When conclusions come before the data, it feels like you are imposing plot. That is true even within a single sentence, as illustrated by examples 8.5 and 8.6.

The important point here is that while it is always essential to *distinguish* results from discussion of them, it isn't critical to *separate* them physically. Each scientific discipline has its own standards for dealing with this, even though most rely on OCAR for the overall story structure. As an author, you decide which approach best fits your field and best serves your story.

8.2.2. Choosing Data to Present

The most important decision in describing results is not *how* to present your data but *which* data to present. We often collect a lot of data, not all of which is needed to build the story. Which to condense? Which to eliminate? These decisions are difficult, both intellectually and emotionally. Deleting data from a paper hurts—it feels like saying your work was wasted. But it wasn't—collecting those data helped you figure out the story and identify the parts that could be cut.

An example of using story to decide how to trim data is a paper that a former student of mine wrote evaluating the seasonal patterns of different forms of plant-available nitrogen in Alaskan tundra soils.[3] Mike Weintraub spent months

3. M. N. Weintraub and J. P. Schimel, "The Seasonal Dynamics of Amino Acids and Other Nutrients in Alaskan Arctic Tundra Soils," *Biogeochemistry*. 73 (2005): 359–80.

developing a high-performance liquid chromatography (HPLC) method to analyze soil amino acids, more months generating the raw data, and still more months
processing those data. Two results were central to guiding the decision of which
data to present: (1) all the amino acids followed the same trends through the year,
and (2) because it is difficult to identify and quantify every amino acid in soil, a
quick and easy colorimetric analysis of total free amino acids (TFAA) was a better
estimate of amino acids as a nitrogen source than the more involved HPLC
method. So was all the HPLC work wasted? Not at all—it was essential. The only
way to find this out was to invest the time and run both methods in parallel. But
this raised the question of how to present the amino acid data. The natural temp-
tation was to present all of it. But that was a huge data set, while the simple story
could be collapsed to two words: "they covaried." The reader didn't need all the
data to get that point; in fact, it would have been a distraction.

Weintraub bit the bullet and collapsed the amino acid data (figure 8.1). The
top panel is critical because it shows concentrations of the three forms of
plant-available N: ammonium (NH_4^+), nitrate (NO_3^-), and TFAA. He considered

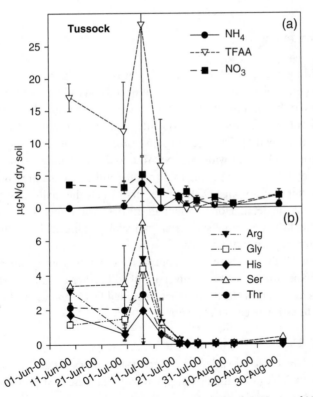

Figure 8.1. Seasonal dynamics of total free amino acids (TFAA), NH_4^+, and NO_3^-
(a), and of the five most common amino acids (b) in tussock tundra soil. This presents
the data for one of four soils analyzed.
Copyright © 2005 Springer. Reprinted with permission.

not showing any data for specific amino acids and saying "all amino acids followed the same trends across time (data not shown)." However, he included the lower panel to more concretely show that you don't need those data and can focus on the TFAA data in the upper panel. Years of work was collapsed into a few subordinate panels in one figure in the paper. Weintraub was able to make the decision to cut so drastically because he knew the story; it was about seasonal patterns of bioavailable N, not only amino acids. Painful, yes, but effective story-telling and effective science writing.

Writers talk about having to "murder your darlings"—the parts you adore but that don't contribute sufficiently to the overall piece. It's good advice for scientists, too.

With our increasing ability to include information in electronic appendices and archives, what you should do in a case like this would be to archive *all* the data. Others may need those data for modeling, meta-analyses, or other kinds of postpublication use. Increasingly, funding agencies are requiring that data be archived and accessible, and journals are beginning to follow suit. So it's a good idea to get in the habit. Too much valuable data has withered away in the filing cabinets of retired faculty.

8.2.3. Presenting Data

After deciding which results to present, you need to figure out how to present them. You could tell us "X was 42," but that would leave the reader wondering why you presented that datum, what it means, and how it fits into the story.

To make it easy for the reader to understand your results, you need give us more than the raw data. You need to synthesize them into a pattern and fit them into the larger story to provide context. You do this by telling a short story about each data set with a clear opening to introduce and frame the presentation.

Most results call for an LD structure: first frame the major point or pattern, then flesh out the detail. Don't present all the details and then synopsize them, or worse, present them without synopsizing or synthesizing at all. Without a framework, readers struggle with details.

Example 8.7 illustrates this LD approach. It comes from a paper that evaluated the factors that regulate the diversity of soil bacteria.[4] The investigators collected soils from across North America, extracted DNA, and used a fingerprinting technique to estimate bacterial diversity.

Example 8.7
Soil bacterial diversity varied across ecosystem types (figure 8.2). Of all soil and site variables examined, soil pH was, by far, the best predictor of soil

4. N. Fierer and R. B. Jackson, "The Diversity and Biogeography of Soil Bacterial Communities," *Proceedings of the National Academy of Sciences* 103 (2006): 626–31. I reproduced only figure 1B, and have deleted text from this paragraph that refers to figure 1A.

bacterial diversity ($r^2 = 0.58$, P < 0.0001) with the lowest levels of diversity observed in acid soils. Because soils with pH levels > 8.5 are rare, it is not clear whether the relationship between bacterial diversity is truly unimodal, as indicated in figure 8.2, or whether diversity simply plateaus in soils with near-neutral pHs. Likewise, because our fingerprinting method underestimates total bacterial diversity, we cannot predict how the absolute diversity of bacteria changes across the pH gradient. When we compare paired sampling locations with similar vegetation and climate but very different soil pHs, we find evidence for the strong correlation between bacterial diversity and soil pH at the local scale. For example, two deciduous forest soils collected in the Duke Forest, North Carolina, showed that the soil with the higher pH (Site DF2, pH = 6.8) had an estimated bacterial richness 60% higher than the more acidic soil (Site DF3, pH = 5.1). Similarly for two tropical forest soils collected <1 km apart in the Peruvian Amazon, the soil with the higher pH (Site PE8, pH = 5.5) had an estimated bacterial richness 26% higher than the more acidic soil (Site PE7, pH = 4.1).

Note how the authors describe this graph; they start with a clear lead—bacterial diversity varied and that variation was driven by pH. Elsewhere in the paper they show the lack of relationship with other environmental variables to reinforce the conclusion that pH is the main control. After making that point,

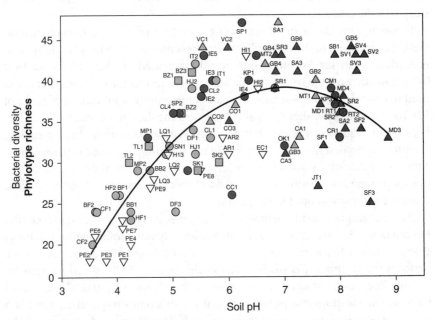

Figure 8.2. The relationship between soil pH and bacterial and phylotype richness, defined as the number of unique phylotypes. Symbols correspond to general ecosystem categories, and labels denote individual soils.

they offer two caveats: they can't tell whether diversity declines at high pH or plateaus, and a fingerprinting technique is not an absolute measure of total bacterial diversity. That done, they go into more detailed analysis of the data and show how more specific comparisons reinforce the overall conclusion—even when soils are close together and have similar vegetation, the more acid soil has lower diversity. This was a very clear LD structure—you get the entire story in the first two sentences, even though the paragraph goes on for an additional five.

8.2.4. Statistics and Stories

In many fields of science, statistics are at the core of data presentation. Statistics allow us to distinguish treatment effects from random variation. Although statistics are essential for establishing the credibility of your conclusions, remember that the story is not in the statistics—it is in the data themselves. When you tell the story through the lens of the statistics, by focusing on the statistical analysis rather than on the data, you steal both clarity and power from the story.

To illustrate telling the story through the lens of the statistics, imagine that you were writing fiction. Would you write "John's affection for Jane was significantly greater than zero"? Of course not, it sounds silly. More important, though, it isn't what we want to know. What we want to know is whether John likes having the occasional cup of coffee with Jane or whether he is desperately, madly, passionately in love with her. Either would mean that he "significantly" likes her, but the story is in the amount.

Similarly, consider a study examining the effect of elevated temperature on methane emissions to the atmosphere, a process important in the climate system. Is it adequate to say "warming significantly increased methane emissions ($p < 0.05$)"? Again, no—the story is in the amount. To model the climate system, you need to know *how much* warming increases emissions. Is it by a factor of 1.2 or of 12? The p-value establishes credibility by describing the data's quality—the difference is not random variation—but it doesn't say anything about their meaning. A better way to describe these data would be "warming increased methane emissions by a factor of 3.4 ($p < 0.05$)." By focusing on the data and making the statistics supporting information, you can tell a story that says more about nature and is more engaging without forgoing rigor.

As an example, consider figure 8.3. In panel A there is a large difference (the treatment is 2.3x the control) that is unquestionably statistically significant. Panel B shows data with the same statistical significance ($p = 0.02$), but the difference between the treatments is smaller. You could describe both of these graphs by saying, "The treatment significantly increased the response ($p = 0.02$)." That would be true, but the stories in panels A and B are different—in panel A, there is a strong effect and in panel B, a weak one. I would describe panel A by saying, "The treatment increased the response by a factor of 2.3 ($p = 0.02$)"; for panel B, I might write, "The treatment increased the response by only 30 percent, but this increase was statistically significant ($p = 0.02$)."

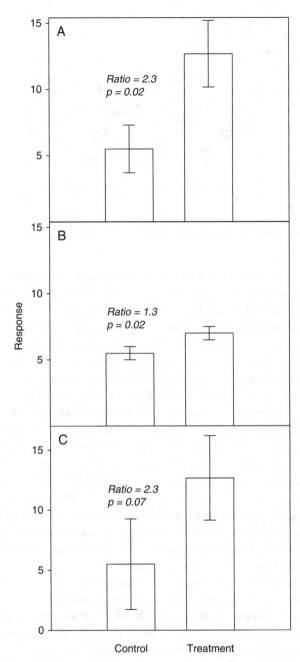

Figure 8.3. Difference versus significance in data presentation.

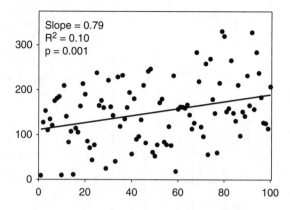

Figure 8.4. Describing the strength of a relationship: fit versus significance.

The tricky question is what to write about panel C. The difference between treatment and control is the same as in panel A (a factor of 2.3), but the data are more variable and so the statistics are weaker, in this case above the threshold that many use to distinguish whether there is a "significant" difference at all. Many would describe this panel by writing, "There was no significant effect of the treatment ($p > 0.05$)." Such a description, however, has several problems.

The first problem is that many readers would infer that there was no *difference* between treatment and control. In fact though, they differed by a factor of 2.3. That is never the "same." Also, with a p value of 0.07, the probability that the effect was due to the experimental treatment is still greater than 90 percent. Thus, a statement like this is probably making a Type II error—rejecting a real effect.

The second problem is that just saying there was no significant effect mixes results and interpretation. When you do a statistical test, the F and p values are *results*. Deciding whether the test is significant is *interpretation*. When you describe the data solely in terms of whether the difference was significant, you present an interpretation of the data as the data, which violates an important principle of science. Any specific threshold for significance is an arbitrary choice with no fundamental basis in either science or statistics.[5]

I might approach panel C by describing the results as they are: "The response in the treatment was 2.3 times higher than in the control ($p = 0.07$)" or, to be conservative, as "The response in the treatment was 2.3 times higher than in the control, but the difference was only significant at $p = 0.07$." These each present all of the information—the difference *and* the probability.

Another example of data versus statistics is in figure 8.4. Is this a strong or weak relationship? If you evaluate these data through the lens of the statistics, you would describe it as a strong relationship because $p = 0.001$: "there is a highly significant positive slope." If you look at the data, however, you would see this relationship as weak—the x variable explains only 10 percent of the variation in y.

5. D. S. Moore and G. P. McCabe, *Introduction to the Practice of Statistics* (Freeman, 2006).

In their own ways, each perspective is correct. This is a robust but weak relationship. The x variable has some influence over the y, but other factors explain most of the variance. The best description of this graph, that is, the most complete and accurate, would be "the relationship between x and y is weak ($R^2 = 0.10$) but statistically significant ($p = 0.001$)."

By focusing on the data, being concrete, and showing the whole story, you effectively and honestly present your results and allow the reader to evaluate them, fulfilling core principles of both writing and science.

8.3. DISCUSSION

Discussion is where you present your thoughts and interpretations, where you answer the questions you posed in the challenge, and where you show your contribution to the larger problem framed in the opening. You have great flexibility in structuring this material, and disciplines differ in their norms for doing it. Giving formulaic advice on how to write the Discussion would therefore be both presumptuous and dangerous. Writing a good Discussion is *the* critical act of creativity in science that no book can teach.

As you sit down to write the Discussion section, however, the first decision you have to make is which internal story structure to use. The Discussion needs to build toward the resolution of the paper, but as a section it should develop a story on its own and have a coherent structure . Some writers use an OCAR structure, opening by reminding readers of the challenge and the question, and then working through to the resolution. Other papers use an LDR structure, opening the Discussion by framing the conclusion—what they showed—and then using the rest of the Discussion to support that argument, building to the overall resolution.

Following are excerpts from several papers; in each example, I present the opening and closing sentences of the discussion. First consider some OCAR examples, where the authors lay out the issue but not the conclusion. The first (example 8.8), is from the field of organic chemistry.

Example 8.8
It is well-known that factors such as the nature of the nucleophile, solvent, and leaving group directly affect the rate of the bimolecular nucleophilic substitution (SN2) reactions; yet, in the case of carbanions, little has been documented with absolute rate constants. . . .

Photoinduced decarboxylation of suitable substituted carbanions provides a route for the formation of substituted cycloalkanes that proceeds in high yields in nonhydroxylic solvents and with good leaving groups such as bromide and iodide.[6]

6. L. Llauger, M. A. Miranda, G. Cosa, and J. C. Scaiano, "Comparative Study of the Reactivities of Substituted 3-(Benzoyl)benzyl Carbanions in Water and in DMSO," *Journal Of Organic Chemistry* 69 (2004): 7066–71.

In this case, the opening of the Discussion reads like a reiteration of the chal-lenge, reminding the reader of what it was, but without posing any conclusion. A similar pattern is illustrated in Example 8.9, which examines how poliovirus enters target cells. Poliovirus has no capsule to fuse with the cell membrane, and the mechanism wasn't clear.

Example 8.9
The extracellular forms of viruses face formidable challenges. The virion itself must be sufficiently stable to protect the viral genome during the pas-sage from host to host and cell to cell, and yet, upon reaching the target cell and encountering the appropriate trigger, the virion must initiate pro-grammed steps that result in the release of the viral genome into the appro-priate compartment of the cell. For nonenveloped viruses, the conceptually simple mechanism of membrane fusion is not an option. . . .

After binding the cell surface, the virus is internalized through a clath-rin-, caveolin-, and flotillin-independent, but actin- and tyrosine kinase–de-pendent, pathway. After internalization (and only after internalization), the virus releases its RNA rapidly from vesicles that are located within 100–200 nm of the plasma membrane without requiring endocytic acidification or microtubule-dependent transport. Our results have settled the long-lasting debate of whether PV [poliovirus] directly breaks the plasma membrane bar-rier or relies on endocytosis to deliver its genome into the cell. These results have also opened interesting questions for this important virus that await further investigation, including what characteristic of these endocytic vesi-cles near the plasma membrane triggers RNA release; and after release near the cell surface, how is the released RNA transported to replication sites.[7]

This Discussion starts by reiterating the problem and reenergizing curiosity before going on to nailing down the answer with the resolution.

Contrast these OCAR Discussions to some that use LDR structure. Example 8.10 illustrates this with a paper about chemotherapy. The study devel-oped a new, nontoxic inhibitor of gamma-glutamyl transpeptidase (GGT), an important enzyme in the development of cancer.

Example 8.10
We have identified a novel class of GGT inhibitors that are not glutamine analogues. Kinetic studies of the lead compound OU749 revealed that the mechanism of inhibition was uncompetitive relative to the γ-glutamyl sub-strate, indicating that the inhibitor bound the enzyme-substrate complex. In contrast to competitive inhibitors, which lose potency as substrate

7. B. Brandenburg, L. Y. Lee, M. Lakadamyali, M. J. Rust, X. Zhuang, and J. M. Hogle, "Imaging Poliovirus Entry in Live Cells," *PLoS Biology* 5 (2007): e183. DOI:10.1371/journal.pbio.0050183.

concentration builds, uncompetitive inhibitors become more potent as the substrate concentration rises in an inhibited open system. . . .

Development of less toxic GGT inhibitors, such as OU749, holds great promise for enhanced cancer therapy.[8]

The first sentence addresses the paper's specific challenge and identifies the important result from the research: a new class of inhibitors. The Discussion elaborates on the properties and benefits of this group of chemicals. This builds to the resolution, which closes the circle back to the paper's opening, which was about developing new chemotherapeutic agents.

The last example is from theoretical physics and offers new developments in string theory.

Example 8.11:
In this paper, we have shown how to construct various D-branes in the type IIB plane-wave background that preserve half the dynamical supersymmetries of the background. . . .

The connection of the instantonic branes we have constructed with instantonic branes in $AdS5 \times S5$ is more obscure. Understanding this could lead to an understanding of the relation between the D-instanton and instanton effects in the dual Yang-Mills field theory. It would also be interesting to understand the effect of the D-instanton on the plane-wave dynamics. Finally, one should be able to analyse the D-instanton contributions by considering the effects of the R4 and related terms in the effective low energy IIB action in this background.[9]

This Discussion starts by identifying the main contribution of the paper: constructing D-branes. It then develops and elaborates that result. The paper resolves by framing interesting new questions that grow from the work. To be honest, I haven't a clue what any of this means, yet the story structure and the message are clear. That is a testament to the writing.

Both OCAR and LDR work well for the Discussion—they each provide a coherent structure that allows you to develop a clear and compelling story. Of the two, LDR is more common; some books even say it is the "right" way to write the Discussion. But that is a rule and so is malleable. With any story you have a choice of structure, and the Discussion should form a story within itself.

8. J. B. King, M. B. West, P. F. Cook, and M. H. Hanigan, "A Novel, Species-Specific Class of Uncompetitive Inhibitors of Gamma-Glutamyl Transpeptidase," *Journal of Biological Chemistry* 284 (2009): 9059–65.

9. M. R. Gaberdiel and M. B. Green, "The D-instanton and Other Supersymmetric D-branes in IIB Plane-Wave String Theory," *Annals of Physics* 307 (2003): 147–94.

EXERCISES

8.1. Analyze published papers

Go through the papers you have been analyzing. Are all the data they presented necessary for the story? Did the authors do a good job of structuring how they present the data—did they use an effective story structure?

8.2. Write a short article

Revisit your short article. Did you use an effective LD structure in presenting the results?

9

The Resolution

Ending well is the best revenge.

Why do we need to get the last word in an argument? Why do Aesop's fables end with a "moral of the story"? Why is the punchline at the end of a joke? The answer to all of these, of course, is because endings are power positions. People remember the last thing you say. The resolution should be your "take-home message," your strongest and most memorable words.

A good resolution shows us how our understanding of nature has advanced, and by offering new insights into the problem identified in the opening, it wraps up the story. This creates the story's spiral, closing back to the original topic, but drawing it out by showing how the starting point has moved. A good resolution achieves this by stepping backward through OCAR: it reiterates the action, answers the questions raised in the challenge, and demonstrates how those answers contribute to the larger problem.

If you put anything but that new insight in the resolution, you undercut it, and with it the entire paper. Because last words are so powerful, people will accept whatever you put there as the take-home message. If you are not careful, some weak or extraneous thought that finds itself in the closing position can come across as your most important.

Don't blow the punchline.

9.1. GOOD RESOLUTIONS

The first example of a good resolution is straightforward, walking backward through the OCAR steps without distraction or complication. The paper examined the enzymes that drive mammalian cells through their cell cycle, focusing on how cyclin–cyclin-dependent kinases become active at specific points in the cycle, driving further development. It examined inhibitor p27, which regulates a kinase complex, and asked whether it has a single mechanism—blocking the enzyme's active site—or whether it also blocks enzyme activation via phosphorylation. The resolution paragraph is as follows.

> Example 9.1
> {1} In conclusion, {2} our data suggest that Y-phosphorylated p27 can inhibit cyclin D-cdk4 complexes by two independent mechanisms: blocking access to the T-loop and disrupting the cdk4 active site directly. {3} Our model suggests that p27 Y phosphorylation is a molecular "switch" that would help turn cdk4 activity on or off. {4} Modulation of Y kinase activity would permit activation of preformed, inactive p27-cyclin D-cdk4 complexes by cdk7 and may be used to regulate cdk4 activity throughout the cell cycle.[1]

This resolution does a number of things well.

{1} The statement "In conclusion" is a flag, telling the reader that what follows is the resolution. Such road signs make it easier to navigate through a paper.

{2} This states that two mechanisms of inhibition are involved. This is the key result of this work, and it answers the question posed in the challenge.

{3} This statement interprets that result and synthesizes it into the idea that p27 Y phosphorylation is a "molecular 'switch.'" That creates a simple message and an accessible intellectual model for how this compound works—switches turning on and off the processes that drive the cell cycle. This starts "widening the hourglass" by moving away from the specifics of how p27 inhibits, to what that means for cell cycle regulation.

{4} This finishes opening the hourglass by bringing the story back to issue the paper opened with—what regulates the cell cycle. It even puts the phrase "the cell cycle" at the end of the concluding sentence, closing the circle back to the opening sentence of the paper, which was: "Cyclin–cyclin-dependent kinase (cyclin-cdk) complexes drive progression through the different phases of the cell cycle by acquiring catalytic activity only at specific points."

1. A. Ray, M. K. James, S. Larochelle, R. P. Fisher, and S. W. Blain, "p27Kip1 Inhibits Cyclin D-Cyclin-Dependent Kinase 4 by Two Independent Modes," *Molecular and Cellular Biology* 29 (2009): 986–99.

This paragraph puts the right ideas in the right places, uses appropriate language to guide you through the paragraph, reiterates the key conclusions and their implications, and closes the circle to complete the story. It isn't elegant or literary in its use of language, but it is inordinately effective in doing its job.

A second example is from materials science and is about producing inorganic/organic composite materials, with a specific focus on semiconducting films that would be useful in electronic devices, including photovoltaics and LEDs. This is a more complex resolution but it achieves the same essential goals.

Example 9.2

{1} These templated nanostructured frameworks thus hold several advantages for the design and synthesis of devices. {2} Films can be selectively deposited through solution phase routes using the chalcogenide affinity to bind to gold. {3} Furthermore, the ability to control the elemental compositions of the nanostructured films allows the band structure of the inorganic framework to be tailored for specific applications. {4} Current research is underway to create composite materials using an organic semiconductor as the structure directing agent. Such materials would make good candidates for device applications such as photovoltaics. Moreover, it is likely that these same band energy trends will hold for nontemplated versions of chalcogenide glass semiconductors synthesized using Zintl cluster precursors. {5} As a result, the data presented here provide a basis to predicatively synthesize a broad range of semiconductors with desired band properties using Zintl cluster precursors and simple solution phase methods.[2]

{1} This is a statement of the overall accomplishment—the authors created a useful material. It gives a clear sense that this is the resolution and that they will flesh out this point in the rest of the paragraph.

{2} In this second sentence, they state the key result from the work: "films could be selectively deposited."

{3} Here they start expanding back out, with a more general interpretation of that result.

{4} The authors continue the widening process, going beyond their specific research and discussing the implications of the work. They expand it by pointing out that "these same band energy trends will hold for nontemplated versions."

{5} Here in their final wrap-up statement, the authors give the most general application of their work: "a basis to predicatively synthesize a broad range of semiconductors."

In contrast to the first example that used the flag words "in conclusion" to tell us we're at the resolution, this uses a whole sentence to frame the point of the

2. S. D. Korlann, A. E. Riley, B. S. Mun, and S. H. Tolbert, "Chemical Tuning of the Electronic Properties of Nanostructured Semiconductor Films Formed through Surfactant Templating of Zintl Cluster," *Journal of Physical Chemistry C*, 113 (2009): 7697–705.

paragraph and guide you through it. It then, however, carries out the same functions—it identifies the key result, opens up the hourglass, and resolves by tying back to the big picture of the paper's opening. This paragraph creates a complete story within itself—opening, developing, and resolving, a strong approach for a longer paper and a longer resolution.

9.1.1. Concluding with a Question

In each of the foregoing examples, the authors identified a result and explained its significance. Sometimes, however, the most important thing you discover is that there is a new question, one you hadn't anticipated, that you want to pose to the community. Fine. Do it, but make the question concrete, and be clear about how it grew from your work—you didn't fail to fill one knowledge gap but identified a new one. Ending with a concrete new question engages a reader's curiosity and can be a powerful way to resolve a paper.

The following is an example of how to resolve effectively with a question. It comes from arctic climate science, examining the mechanisms responsible for the enormous loss of sea ice in the Arctic Ocean during 2007. That loss shocked the climate community—such massive ice loss had been predicted to be decades away.

Example 9.3
{1} There was an extraordinarily large amount of ice bottom melting in the Beaufort Sea region in the summer of 2007. Solar radiation absorbed in the upper ocean provided more than adequate heat for this melting. An increase in the open water fraction resulted in a 500% positive anomaly in solar heat input to the upper ocean, triggering an ice–albedo feedback and contributing to the accelerating ice retreat. The melting in the Beaufort Sea has elements of a classic ice–albedo feedback signature: more open water leads to more solar heat absorbed, which results in more melting and more open water. The positive ice–albedo feedback can accelerate the observed reduction in Arctic sea ice. {2} Questions remain regarding how widespread this extreme bottom melting was, what initially triggered the increase in area of open water, and what the summer of 2007 portends for 2008 and beyond.[3]

{1} The main part of this resolution states both the findings and conclusions using clear strong language: "solar radiation . . . provided more than adequate heat for the melting" and "The positive ice–albedo feedback can accelerate the observed reduction in Arctic sea ice." There is no hesitation or weakness.

3. D. K. Perovich, J. A. Richter-Menge, and K. F. Jones, "Sunlight, Water, and Ice: Extreme Arctic Sea Ice Melt during the Summer of 2007," *Geophysical Research Letters* 35 (2008): L11501. DOI:10.1029/2008GL034007.

{2} The resolution goes further to frame a series of questions about both the mechanisms involved and the implications for the future. However, rather than undermining the conclusions, these questions actually reinforce and extend them; they point the direction forward. They engage a reader's interest. Even using a word like *portend* emphasizes the new question—it's an ominous word.

Here is a slightly more complex example from astrophysics.

Example 9.4
{1} Finally, while the details of the solutions that we have discussed specifically apply to the case of a rotating NS accreting from a disk fueled by a companion star, the general feature of a multiplicity of states available for a given mass inflow rate of matter can probably be generalized to other accreting systems in which recycling occurs. {2} An example is that of an accretion disk around a rotating black hole. Numerical simulations show that while a fraction of the accreting mass is ejected through a jet, another fraction, of slower velocity and at larger angles from the jet axis, falls back into the disk, getting recycled. {3} It would be interesting to include this mass feedback process into numerical simulations of accretion disks around black holes and to investigate whether the discontinuous states and cyclic behavior might ensue in those cases as well.[4]

{1} Here, the authors start with the specific results of the work, defining the system it is limited to, but start opening the hourglass back up by suggesting that this multiplicity of states "can be generalized."
{2} They develop that by highlighting a particular system that they think it would generalize to the black hole's accretion disk and how it would fit into their analytical framework.
{3} Finally the authors frame the new question that grows from their work and pose it to the community. It's an interesting question and in no way undermines the accomplishment of what the authors did. Rather, it creates a natural progression. This is nicely done.

9.2. BAD RESOLUTIONS

All of the previous examples illustrate strong resolutions, and, despite the different writing approaches, they all carry out the core functions of identifying the main results and their implications. Now let's consider bad ones—resolutions that fail in those core functions. There are several ways to destroy a good paper with a

4. R. E. Bozzo Perna and L. Stella, "On the Spin-up/Spin-down Transitions in Accreting X-Ray Binaries," *Astrophysical Journal* 639 (2006): 363–76.

bad resolution. You can be weak, distracting, or, at worst, you can actively under-mine your conclusions.

9.2.1. Weak

Weak resolutions fail to frame the conclusions. In this type of ending, authors usually synopsize their results and then tell you that they are important, but don't clarify how—they don't answer the questions they were asking and don't synthe-size their information into knowledge. Here's an example.

> Example 9.5
> A proteomic evaluation of hummingbirds under simulated migratory condi-tions revealed evidence of several stress-associated processes: protein degra-dation in wing muscle tissues, depletion of metabolic cofactors, and enhancement of stress-response proteins. These results suggest that changes in the hummingbird proteome may provide new insights into the complex physiology of avian systems biology.

This paragraph does a good job of synopsizing the results, but then it stumbles. Rather than synthesizing new knowledge, it skips that step. Instead, it simply tells us that the research is important and has implications beyond hummingbirds. In doing so, it overreaches and underdelivers, being simultaneously unconvincing and obvious.

These authors were trying to widen the hourglass to reach the largest possible audience, which is commendable, but they did it badly. They tell us that it "may provide new insights . . . into avian systems biology," but they don't tell us what those insights are! What did these authors contribute to the wider field of bird physiology and ecology? We're left to figure it out for ourselves. We might con-clude that the authors didn't fully understand their own data and are tossing them out in the hopes that we'll figure it out for them. That isn't the take-home message you want to give readers.

As to the work's implications to the wider field, it goes without saying. Would a scholar studying migration in geese, albatrosses, or swallows ignore a paper on hummingbirds? No. Saying it is relevant accomplishes nothing without the concrete substance to illustrate that relevance. This is a train wreck of a resolution.

To fix a resolution like this, you need to identify the new insights.

> *"A proteomic evaluation of hummingbirds under simulated migratory condi-tions revealed evidence of several stress-associated processes: protein degrada-tion in wing muscle tissues, depletion of metabolic cofactors, and enhancement of stress-response proteins. While hummingbirds migrate long distances over water without feeding or resting, it is physiologically stressful, and the birds' ability to manage this stress may limit the distance they can migrate."*

Here, rather than trying to make a methodological but largely meaningless suggestion about how to study birds in general, the paper ends with a clear conclusion about what these data mean—migrating is stressful—and a suggestion for what they say about hummingbird biology and behavior, suggestions that clearly relate to other birds. This resolution says something concrete—it resolves.

If the authors wanted to open the hourglass wider to explicitly encompass other migratory birds, they could modify the last sentence to make humming-birds a member of that larger group:

> "While many birds, such as hummingbirds, migrate long distances without feeding or resting, it is physiologically stressful, and birds' ability to manage such stress may limit the distance they can migrate."

This adds "such as hummingbirds" to make it clear that they are an example; it also condenses "and *the* birds'" to "and birds'," a subtle change that shifts it from referring to specific birds to birds in general. This does, however, suggest that we know migrating is stressful in other birds and that we had developed that argument in the Introduction. If that isn't the case, this might be an excessive stretch to draw from this study alone.

Here's another example of a weak resolution, this one from the field of medical microbiology.

Example 9.6
In summary, we show that X7 alters the expression pattern of extracellular proteases in the "flesh-eating bacterium" Streptococcus pyogenes, which causes necrotizing fasciitis. If the function of X7 can be fully established, it would likely deepen our understanding of this destructive disease.

In this example, the authors clearly knew they couldn't end with a summary of the data; they needed some kind of wrap-up. But all they could come up with was a throw-away line that included a patently obvious truth and a back-handed slap at their own data. Of course if we understand what controls the expression of protein-destroying enzymes, we would understand the disease better! And they remind us in the last sentence that they haven't fully established the function of X7. There's nothing wrong with not fully establishing its function, but don't end a paper by telling your readers what you didn't achieve.

As in the previous example, the fix is to make a concrete conclusion that synthesizes the results into knowledge and provides a meaningful take-home message.

> "In summary, we show that X7 alters the expression pattern of extracellular proteases in the 'flesh-eating bacterium' Streptococcus pyogenes, which causes necrotizing fasciitis. This research may offer a route to developing therapeutic agents that would minimize tissue damage while antibiotic treatments were directly attacking the bacterium itself."

This version identifies the work's real conclusion—it might lead to drugs that would provide a tool for managing the disease. That would close the circle to the paper's opening, which framed a story about necrotizing fasciitis, the bacterium that causes it, and potential therapies for a horrible malady.

Note that this resolution, while ending with a strong message, is carefully constrained. It doesn't say that X7 is necessarily going to be that new therapeutic agent, and it only says that this "may" offer a route—it might not work in vivo. You can make a strong statement without overselling.

9.2.2. Distracting

Some papers conclude with material that is distracting—ideas that should be in the Introduction or is already in textbooks and that neither synopsizes nor synthesizes the results. The next example is from a paper about forest tree nutrition, asking how much organic N is taken up by mycorrhizal fungi, which acquire nutrients from the soil and transport them to the root.

Example 9.7
The mycorrhizal fungal hyphae extending out from tree roots can comprise more than 1/3 of the total biomass of microbes in the soil. They greatly extend the absorptive surface area of the root system and enhance total nutrient uptake by the trees. Additional work, however, is required to assess how much mycorrhizal fungi enhance the uptake of organic N forms in forest soils.

These first two sentences are truisms that have been known for decades—textbook material, rather than results of this particular study. The only thing this paragraph says about the study itself is that additional work is required to assess how much organic N mycorrhizae take up. So did we learn anything? In the paper we actually did, but not from this resolution—it resolves nothing and merely distracts from the story.

A second way a resolution can be distracting is by introducing new information at the end. The following might appear to be a strong resolution.

Example 9.8:
In arid environments such as East Africa, termites are critical "ecosystem engineers." They collect resources such as nitrogen and phosphorus from far afield and accumulate it in and near their mounds, creating nutrient hot-spots on the landscape. These hot-spots may be sites for colonization by new seedlings of both the native savanna trees and for novel invasive plant species.

The problem here is that invasive plants were never mentioned in the Introduction. The idea that termite mounds create invasion sites is interesting and important, but it *must not* be a new idea, first raised in the resolution.

The resolution *must* close the circle back to the opening. Instead of closing the circle, however, this resolution goes haring off in a new direction.

I would guess that the idea of termite mounds creating invasion sites developed while the authors were writing the paper. That's great; developing new ideas while you are writing is exactly what Montgomery meant when he said that "clear thinking can emerge from clear writing." Never close your mind to new insights about your work and its implications. But when you have them, go back and weave them into the opening and Introduction. You are not writing a "whodunit" mystery where shocking plot twists are expected. You are writing science, where such plot twists are forbidden.

From the perspective of getting your message out to the widest possible audience, surprise resolutions are a disaster. The points they make won't get picked up in a literature search, so potential readers will only find your paper by accident. Plant ecologists should know about work showing that termites facilitate invasion, but they probably wouldn't learn it from this paper. Everyone loses—plant ecologists miss useful information, and the authors lose citations.

9.2.3. Undermining Your Conclusions

The worst possible way to end a paper is to actively undermine your conclusions, and yet this may be the most common way to end scientific papers. Many end by saying "more research is needed to clarify our findings." Resolving a paper this way focuses on what you *haven't* accomplished. That is worse than throwing away a power position—it uses that power to weaken your conclusions and your science.

I understand the humility involved in "more research is needed"—we know our work isn't perfect and that there are still questions about both the big issue of our opening and the small issue of our challenge. Uncertainties remain. But the resolution is not the place to discuss them.

A really egregious example of undermining the conclusions is in the next example.

Example 9.9
To conclude, 3-methyl-ambrosia offers a new approach for thyroid carcinoma therapy. Our data provide evidence on safety and in vivo activity of this compound in patients with this condition, although the proof for clinical benefit remains to be established in future clinical trials.

In the first part of this passage, the authors tell us that they have a new therapy that appears safe and effective. Their take-home message, though, is that they don't know whether it really works! Talk about destroying the story. This would have been much better as:

"*While further clinical trials will be necessary to establish the full benefits of 3-methyl-ambrosia as a therapeutic agent, our data provide evidence that it is safe and shows in vivo activity against thyroid tumors. 3-Methyl-ambrosia therefore may offer a new approach for treating patients with thyroid carcinoma."*

This version says the same things as the original, but strongly and positively. It is clear that additional clinical trials are necessary and that the efficacy remains uncertain (it only "may offer" a new approach). But it ends by highlighting the authors' intended message: 3-methyl-ambrosia may be an effective new anticancer drug.

The earlier example about flesh-eating bacteria also undermined itself with a "more research is needed" expression as well: "if the function of X7 can be fully established."

This kind of "more research is needed to clarify our results" statement is fundamentally different from "concluding with a question" that I illustrated in examples 9.3 and 9.4. Those resolved by posing concrete questions that grew *from* the work. "More research is needed" poses questions *about* the work and makes it sound like the author didn't complete the research.

There are many ways to undermine your results, including expressions such as "but the importance of this has yet to be assessed," "we hope that this review will simulate further research to answer the many unanswered questions," "this topic deserves more research," and so on. All of these use fuzzy expressions that suggest weaknesses in the existing work, rather than expressing substantive conclusions or pointing out clear new questions.

9.3. HOW TO FIX A BAD RESOLUTION

Although the problems in the bad examples are superficially different, they share the same core problem and solution. In each example, the authors ended with a line that concludes little, weakens the existing conclusions, or fails to complete the circle of the story. Each version misses some component of the SUCCES formula—frequently C, concrete.

The solution is to first pare away the dead tissue—the fluff, the detractions, and the new ideas. When those are important, move them elsewhere. Then, condense your resolution to do three things: (1) synopsize the key results, (2) synthesize those results—show us how they answer your question, and (3) show us what this contributes to solving the larger problem. If you achieve those three objectives, each clearly and concretely, you will have a strong resolution that ends your paper with maximum punch.

9.4. RESOLUTIONS IN PROPOSALS

Most proposal authors recognize that they need to grab a reader's attention quickly—the critical energy is right up front in the initial action or lead. But many go all the way to a pure LD structure. After describing the proposed experiments, they end, having said everything that seems to need saying. There is no synthesis, no wrap-up, no resolution.

That's a mistake. Make space for a resolution paragraph that encapsulates the proposal, reiteratesthe big issue and explains how the components work together to address it—make the final pitch for why the proposal should be funded. Some may argue that they have already made those points, so repeating them would waste valuable space.

But reviewers' thoughts may not be fully crystallized. They have just worked through pages of dense, detailed material. They read about multiple hypotheses and lines of experimentation. They are thinking about how the pieces fit together, whether the experiments will work, whether this will really solve the problem, and, importantly, what to write in their review. This is your last opportunity to give them the words. To convince them to check "excellent;" or as the program officer holds your fate in her hand, hovering over the line on the whiteboard, asking "which side does this go on?," to say "must fund."

To illustrate a proposal resolution, here is one from a proposal evaluating why nitrogen availability in arctic tundra soils crashes in the middle of the growing season and how that affects overall ecosystem function and C-cycling.

Example 9.11

The Arctic tundra is one of the world's major stores of C and the possibility that the temperature-decomposition-CO_2 flux positive feedback may accelerate climate warming is a concern in Arctic and global climate studies. The "spin-off" feedbacks via nutrient effects on vegetation change may further accelerate the climate-warming feedback. However, biogeochemical models assume that decomposition is limited by C-availability but regulated by temperature. Thus, the assumption is that if temperatures rise and the snow-free season lengthens, decomposition and CO_2 release will increase dramatically. However, C cycling is N-limited in tundra, at least later in the growing season after the nutrient crash, challenging these ideas. We propose research that will span multiple scales to evaluate the mechanisms causing the "nutrient crash," how they are driven by seasonal weather patterns and plant phenology, and what the effects of the nutrient crash on C-cycling will be. This work will require intense mechanistic work focusing on transitions and transformations that occur over only a few weeks at most, but which have profound impacts on the tundra ecosystem. We will scale this mechanistic work to the intermediate spatial scale by doing transect measurements along the Kuparuk Basin to validate that patterns that occur locally are robust. We will scale to the whole Arctic system by integrating these mechanisms, and importantly the N-effects on decomposition, into the MEL model that is designed to explore multiple limiting resource effects on ecosystem function. As an integrated package, this research will explore how the changing seasonal pattern that drives the crash in nutrient availability in tundra soils will alter overall tundra C-cycling and its role as a source or sink of C and through this its role in the global climate system.[5]

This paragraph reiterates the entire proposal, laying out the problem, challenge, research, and how it will solve the problem. This takes a half-page, which seems like a lot when you're struggling with a page limit. But the feedback from the program officer suggested that this barely squeezed over the line into the "funded" category. We'll never know whether that resolution paragraph was what gave it the million-dollar nudge, but I do know that a reviewer's opinion is sometimes not solidified until the end. So end strong.

EXERCISES

9.1. Analyze published papers

Examine the resolutions of the papers you are evaluating. Do they effectively resolve? Do they briefly sum up the most important results? Do they answer the question? Do they close the circle by returning to the big problem identified in the opening? If not, how would you rewrite the resolution to achieve these goals?

9.2. Write a short article

Do the same exercise for the short articles you and your peers are writing.

5. From M. Weintraub, lead PI, "The Changing Seasonality of Tundra Nutrient Cycling: Implications for Ecosystem and Arctic System Functioning," funded by the U.S. National Science Foundation's Arctic System Science Program.

10

Internal Structure

I figured out, over and over, point A, where the chapter began, and point B, where it ended, and what needed to happen to get my people from A to B.
—ANNE LAMOTT, *Bird by Bird*

OCAR defines the overall structure of a story. The opening grabs your attention with characters and a setting that you care about. The challenge creates uncertainty and curiosity: what is going to happen to those characters? Novelists describe that as creating "tension"—the emotional drive to keep reading. The action feeds you information and develops the story. Finally, the resolution rewards your efforts and relieves the tension—our hero and heroine finally get together, our questions are answered! We may not feel the emotional intensity in a science paper that we do in a good novel, but the tension that keeps us reading is fundamentally the same—it's grounded in curiosity. We don't bother reading a paper if we already know the story. This flow of opening, development, and resolution—building and then rewarding curiosity—creates a story's "arc" (figure 10.1). The vision of story as arc also emerges from the idea that a story has a spiral structure, moving forward but coming back, at the end, to where it began.

Scientific writing is successful when it creates that flow and that arc. But papers and proposals are made up of sections, each of which tells its own story and has

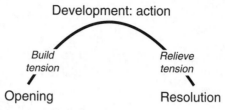

Figure 10.1. A story arc.

its own arc. The Introduction tells us why you did the work—it opens, narrows, and resolves with the paper's overall challenge. The Materials and Methods starts with the study system, then the measurements, and wraps up with how you analyzed the data. The Discussion, too, should form a story of its own, as I argued in chapter 9. It opens by restating the issue, discusses the evidence, and resolves with the paper's conclusion.

These major sections, however, are built of discrete modules: subsections that describe a single method, a single data set, or a single argument. Those subsections should be written to package complete ideas—that is, form story arcs of their own. When you describe a method, you should tell us what information you were trying to gain and what you did to get it. In describing a result, give the overview, the specifics, and the significance.

Going further, each subsection is built of units finer still: individual paragraphs, sentences, and clauses within a sentence. Each tells its own story and has its own structure—they should each form an arc. A story, therefore, doesn't have just a single overall arc, but a hierarchical structure, with small arcs nested within larger ones, ultimately creating the whole (figure 10.2). This hierarchy is analogous to the structure of matter: quarks within protons within atoms within molecules.

A good story works when this hierarchical structure works. Each little arc draws readers forward—it grabs them with a local opening, engages them with a snippet of action, and then rewards them with a resolution. Each forms one turn of the spiral, and step-by-step, carries the reader from the initial issue to the final resolution.

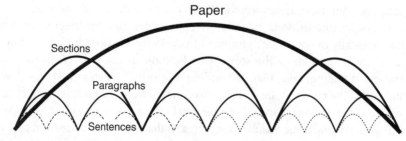

Figure 10.2. A story is a set of nested arcs.

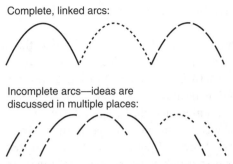

Figure 10.3. Complete versus broken story arcs: beginnings and endings are power positions.

Creating arcs compartmentalizes your thoughts and makes them manageable. It works because we learn by placing information into an established framework. For each new point we build a structure: we give the context and then describe the information, which makes a small story arc. That provides context for the next point, allowing us to construct increasingly complex stories one piece at a time. This is like computer programming. Good programs are built from subroutines or objects, each of which is internally complete; they interact with others by passing specific pieces of *processed* information. Bad programs look like plates of spaghetti. Writing is linear, so we have to lay out these modules in series, each building from the previous.

Effective arcs make it easier for readers to deal with multiple ideas in a single paper. Compare the visual patterns when you link three complete arcs and when you break them up and intersperse them (figure 10.3). The two patterns each contain three arcs and take the same amount of space, but the bottom one is painful to look at (it makes me dizzy). When you write that way, it's just as painful.

Arc structure is effective as well because beginnings and endings are power positions. They emphasize the information contained there, saying *this* is an important point. Without such positions, readers have a hard time distinguishing what is more important from what is less. Creating discrete arcs creates and highlights those power positions. In the top panel of figure 10.3, there are four: the beginning, the end, and the two connecting points in the middle. The bottom panel loses those points; even the beginning and ending are muddled.

10.1. EFFECTIVE ARCS

The last six chapters were all about how to build effective story arcs. They focused on the entire paper, with its opening, action, and resolution. The same principles, however, apply at every level of organization. I've discussed at length how to create the arc of the Introduction (chapters 5–7) and touched on the Discussion (chapter 8). Here I illustrate how to use the same principles to write effective subsections before going on to even finer scales—paragraphs and sentences—in the following chapters.

Let's look at some examples of sections from papers that make effective story arcs. The first is from a paper that evaluated patterns of nitrogen retention in watersheds in the Eastern United States. Nitrogen enters these watersheds from atmospheric deposition, fertilizer, and sewage treatment plants. It runs into rivers as nitrate (NO_3^-) and is ultimately flushed into the ocean. The authors found that in southern watersheds, less of the N entering them reached the ocean; these watersheds "retained" more N. They analyzed the factors that caused this pattern and developed a hypothesis that the warmer ecosystems in the South allowed more biological denitrification, in which NO_3^- is converted to gaseous products: nitrous oxide (N_2O) and N_2. The following section develops one piece of that overall story, evaluating the effect of human population on the proportion of N retained versus exported. I flagged important points with numbers in curly brackets.

Example 10.1
{1} Population density
{2} Increased N export has been shown to be related to increased human population density. {3} If northern watersheds had higher population densities than more southern systems, it is possible that the attendant increase in N input could potentially result in less watershed processing and hence an increase in proportionate riverine export in northern systems. {4} However, the ranges of population densities in northern and southern watersheds overlapped considerably, and, for a given density, the proportion of N export was always higher in northern systems than in southern ones. When all watersheds were considered together, there was a significant relationship between population density and riverine export due only to the influence of the three most densely populated watersheds (the Charles, Blackstone, and Schuylkill). {5} We conclude that population density cannot explain the difference in N export between northern and southern watersheds, but it is a factor to consider for explaining high N export in some watersheds.[1]

{1} The passage starts with a subhead to identify the issue: human population density.
{2} This is the opening (O), which identifies the topic and characters: N export and human population density.
{3} This poses the challenge (C): if northern watersheds have more people they might export more N.

1. S. C. Schaeffer and M. Alber, "Temperature Controls a Latitudinal Gradient in the Proportion of Watershed Nitrogen Exported to Coastal Ecosystems," *Biogeochemistry* 85 (2007):333–46.

{4} Here, the action (A) starts. The authors briefly analyze the data before coming to the climax: the apparent relationship between export and population density was driven by just three rivers.

{5} The story resolves (R) with the statement "We conclude that population density cannot explain the difference in N export between northern and southern watersheds." The phrase "We conclude" is a flag that this is a resolution point.

This passage contains an entire OCAR story arc that completes the discussion of human population density—this is the last time the paper discuses it. After this, the authors move on to analyze other factors that might explain greater N retention in southern watersheds. This is an effective use of story arc structure.

A second example is from a paper that is also about global N cycling, but which asks the question: "What processes cause N to be in short supply in many terrestrial ecosystems?" When the atmosphere is 78 percent nitrogen (N_2) and bacteria can convert that N into biologically available forms, why are plants limited by N? The specific section discusses N relative to carbon (C) and how organisms have characteristic ratios of those elements in their tissues.

Example 10.2

Stoichiometry

Organisms use essential elements at characteristic ratios, and these ratios differ systematically among different groups of organisms. Element ratios are widely used in the analysis of marine ecosystems. Their application is usually less explicit in terrestrial ecology, but they provide the basis for using critical element ratios to predict element mineralization or immobilization during decomposition. One general feature of terrestrial ecosystems is that C:element ratios in plants, especially trees, are much wider than those in other organisms as a consequence of plants' use of C-based compounds (cellulose, lignin) to provide structure. For N in particular, soil bacteria generally have a C:N ratio near 6, while plants often have C:N ratios > 100. Even the leaf litter produced in forests on infertile soils can have C:N ratios in excess of 100.

Consequently, relative to their own requirements, animals and microbes live in a C-rich, N-poor world. Animal nutrition and growth are often constrained by the N content of their food, and protein deficiency is widespread. This difference in stoichiometry can sustain N limitation to animals even where plants are not limited by N supply. Microbes also encounter little N (relative to their requirements) in the plant litter they decompose, and so they retain the N they obtain from their substrate and acquire more directly from inorganic pools in the soil. As a result, N cycling from organic matter back to biologically available forms lags behind the decomposition of plant litter.[2]

2. P. M. Vitousek, S. Hättenschwiler, L. Olander, and S. Allison, "Nitrogen and Nature," *Ambio* 31 (2002): 97–101.

This example opens by defining the overall issue and characters (organisms, essential elements, and characteristic ratios). The two paragraphs discuss these ratios and how they vary among different groups of organisms. It resolves by telling us about why organisms are N-limited: "As a result, N cycling from organic matter back to biologically available forms lags behind the decomposition of plant litter." This section tells a discrete story, one that helps build the larger argument.

What sets this passage off from example 10.1 is that these authors split the story into two paragraphs, each of which makes its own arc. The first paragraph develops the idea that all organisms have characteristic element ratios, but it focuses on plants. That focus is established in the first sentence: "One general feature of terrestrial ecosystems is that C:element ratios in plants, especially trees, are much wider than those in other organisms." Although one sentence starts by mentioning bacteria, it ends by saying "while plants often have C:N ratios > 100," returning the stress to plants. The last sentence of the paragraph starts with "Even the leaf litter," making plants the sentence's subject, and so closes the story arc about plants.

The second paragraph, in contrast, opens with "Consequently, relative to their own requirements, animals and microbes," which makes animals and microbes, the organisms that *consume* plants, the prime characters. All the sentences in this paragraph are about plant consumers, rather than about plants. Thus, it makes a story of its own.

Together, these paragraphs explain why organismal stoichiometry regulates N flow through an ecosystem and why N limits plant growth. The authors could have written this as one long paragraph, but it is stronger this way—each paragraph forms an independent arc. They linked them together with the word *consequently* and in the second paragraph by picking up the idea that the environment is C-rich and N-poor.

Through the devices of subheads, paragraph breaks, and flag words such as *however* and *consequently*, successful authors guide the reader through story arcs and arguments. Individual arcs integrate to form the overall paper. Watching your arcs and ensuring they are coherent and connected gives structure and flow to your writing. Making and resolving complete story arcs makes the reader's job easy.

10.2. ARCLESS WRITING

When writing lacks clear story arcs, it becomes an incoherent mass with no obvious direction, no internal structure, and no points of clear emphasis. The reader may learn little from the work, or worse, they may misinterpret it. Example 10.3 illustrates such arcless, and artless, writing.

Example 10.3
California supports rich fisheries off its coast. The high productivity of fish is supported by high rates of algal production. Algal growth in the ocean is

typically limited by nitrogen supply, but this is high off California because N-rich deep water wells up to the surface along the coast. This upwelling is driven by winds that push the south-flowing surface water away from the shore, allowing deep water to rise to the surface. These off-shore winds are driven by regional climate patterns, including El Niño, that are being intensified by the greenhouse effect, which results from increased CO_2 in the atmosphere. Increased CO_2 in the atmosphere also increases the amount of CO_2 dissolved in the ocean, which reacts with water to form carbonic acid (H_2CO_3), reducing the ocean's pH. This reduced pH makes it hard for shell-forming organisms to make calcium carbonate shells, and so may reduce the productivity of important marine species such as abalone, oysters, and even sea urchins. Thus, increasing atmospheric CO_2 is going to have many important effects on marine ecosystems.

I find this paragraph completely incoherent, but I wrote it to be that way. Importantly, though, it isn't incoherent because the sentences are unclear; they aren't—each is a simple declarative statement. It also isn't incoherent because the sentences don't link together. Rather, the opposite is true: each sentence builds tightly off the idea developed in the previous one. Look at the ideas each sentence starts and ends with. They tie together seamlessly.

California . . . fisheries
Fish . . . algae
Algae . . . nitrogen . . . upwelling
Upwelling . . . winds
Winds . . . climate
Climate . . . greenhouse effect . . . CO_2
CO_2 . . . acid . . . reduced pH
Reduced pH . . . damage to shell-forming organisms
Thus, . . . CO_2 will affect marine ecosystems.

The problem is that although the sentences flow, they don't flow anywhere in particular. This feels like the result of a game where the first person writes a sentence, passes it to the next person, who writes one and passes it on to the next. The paragraph opens by identifying characters of California and its fisheries, but then keeps adding new characters and new directions—first nitrogen and upwelling, winds and climate, CO_2, acidification, shellfish, and finally marine ecosystems. It then ends with a resolution statement that, although true, has no closure back to the opening.

This paragraph lacked thematic coherence. As a result, the story was unclear. Is it about how climate change will affect California fisheries? Or is the fisheries example intended to illustrate the larger theme of climate effects on oceans?

The paragraph is incoherent because it fails to develop the OCAR functions into an effective structure. There is no clear point and no arc to the story. It drifts. This kind of writing can emerge because the authors never knew where they were going, or because they got distracted in the middle by CO_2 and allowed the story to float off into uncharted and unplanned territory. In revising a paragraph like this one, you need to figure out what the story arcs are and break them into separate units.

Example 10.4
California supports rich fisheries off its coast. The high productivity of fish is supported by high rates of algal production. Algal growth in the ocean is typically limited by nitrogen supply, and is high off the California coast because N-rich deep water wells up to the surface along the coast. This upwelling is driven by winds that push the south-flowing surface water away from the shore, allowing deep water to rise to the surface. These winds are driven by regional climate patterns, including El Niño, that are being intensified by the greenhouse effect. Thus, the productivity of California fisheries will likely change as a result of climate warming, and the changes may result via complex and unexpected mechanisms such as changes in ocean circulation patterns.

In addition, increasing CO_2 is causing the pH of the ocean to decline, and this may have separate but important effects on California fisheries. As CO_2 increases in the atmosphere, more dissolves into the ocean as carbonic acid (H_2CO_3) . . .

Now this passage is structured in coherent arcs—the first, main one, is about how fishery productivity is driven by ocean circulation and thus is sensitive to climate change. Then I pulled the information about acidification into a separate arc that follows but connects to the first one through the climate change-CO_2 link and the language "in addition." By creating linked arcs, the writing gained coherence and SUCCES-type "simplicity." Instead of being a complex mass of interwoven information, it now has a series of simple messages that together add up to a larger, equally simple message: increasing atmospheric CO_2 is going to alter California fisheries.

This kind of arcless writing usually results from what I call "stream of consciousness" writing, in which the author puts down each thought as they come to mind, one idea stimulating another and all flowing onto the page. Many inexperienced authors write this way. Undergraduate essays written the night before the deadline are notorious for this, with no visible structure, ideas appearing in multiple places, and incomplete arguments. Even experienced writers find extraneous thoughts inserting themselves—one sentence sparks a thought, and into the paragraph it goes. What distinguishes an experienced writer is that those extraneous thoughts don't survive to the final draft. If they are interesting and germane, they go into their own arc elsewhere, otherwise they go in the trash.

To ensure that your final pieces have an effective internal structure, go over them paragraph by paragraph and section by section and ask the following questions:

Does each unit make a single, clear point?
When several paragraphs together form a section, are the linkages among them clear?

Has every extraneous thought that breaks the serial arc structure been removed?

When you introduce a topic, do you resolve that discussion before introducing a new topic?

Is every major unit of the work defined by either a subhead or clear opening text?

If you can't answer "Yes" to each of these questions, then you haven't finished working on the structure.

EXERCISES

10.1. Analyze published papers

Go back, once again, to the papers you've been analyzing. This time look at their internal structure. Can you block out sections that form complete arcs? How do the authors indicate that they are beginning or ending arcs? Identify the theme of each arc and give it a subheader that describes that theme.

10.2. Write a short article

Go back again to your short article. Evaluate your own story arcs. Do you form complete arcs, or do you have ideas that keep cropping up in multiple places? Rewrite to ensure clear, well-defined arcs.

11

Paragraphs

Make the paragraph the unit of composition.
—STRUNK AND WHITE, *Elements of Style*

Words are to sentences what atoms are to molecules: the basic building blocks that control structure and function. If we extend that analogy, paragraphs become cells: the fundamental unit of life. A cell gains life from its structure, a structure that creates internal cohesion and external connection, allowing it to function as part of a larger organism. So, too, with paragraphs. Hence, Strunk and White's second principle of composition: "Make the paragraph the unit of composition."

But how do you make a paragraph a cell, or a unit of composition? What do those terms even mean? Paragraphs tell stories. Not surprisingly, therefore, a paragraph becomes a unit of composition when it tells a complete short story with a coherent structure, a story that fits into and contributes to the larger work. If you string sentences together until you need to come up for air, and then throw in a paragraph break, you will not have a unit of composition.

Grade school teaches that a paragraph has a topic sentence that makes a point, which the rest of the paragraph develops. This topic sentence–development (TS-D) model of paragraph structure is a simplified version of the lead/development (LD) story structure I introduced in chapter 4. It works much of the time and is a good starting point. However, part of all advanced training is unlearning the simplifications you were taught in introductory classes. Those classes build simple schemas to get you started in the field, but to advance you must move beyond them. Electrons don't orbit around the nucleus like planets around the sun, single genes don't necessarily code for single proteins, and paragraphs don't necessarily have a TS-D structure.

You can write a paragraph using any of the story structures discussed in this book. Each paragraph needs an opening that sets the stage, and each needs to resolve by making a point, but those don't have to be either a single sentence or the first sentence. For choosing a structure, the most important decision is whether to make a point and then develop it, producing an LD structure, or to build to a conclusion, producing an OCAR or LDR structure. Joseph Williams[1] distinguishes these as "point-first" versus "point-last" paragraphs.

11.1. POINT-FIRST PARAGRAPHS

The simplest form of point-first paragraph is the classical TS-D structure. If a paper were written with only TS-D paragraphs, you could skip along, reading the first sentence of each paragraph, and still get its essence. TS-D is simple, clean, and works well for most jobs. It should dominate your writing. If you go back to chapter 8 (see examples 8.1 and 8.4), you'll note that this was how I suggested describing your methods and results. Here are several other simple TS-D paragraphs.

Example 11.1
A Result: Neither calculation reproduces the experimental strength distribution. The distribution for GXPF1A is closer to the data, but it pushes the strength up too high in excitation energy. An even more dramatic increase occurs for the calculation with KB3G, although the strength integrated up to 7.5 MeV reproduces the experimental value quite well. The summed B(GT) strength up to Ex = 7.5 MeV (a total of 48 states) for the KB3G interaction is $\Sigma B(GT)KB3G = 2.02$ (with a further 10% of that value located at energies up to 10.3 MeV) compared to the experimental value of 1.95 ± 0.14 up to that excitation energy. The summed strength up to *Ex* = 7.5 MeV with the GXPF1A interaction is $\Sigma B(GT)_{GXPF1A} = 2.65$. A further 8% of that value is located at higher excitation energies, fragmented over many weak states.[2]

1. Joseph Williams 1981. Style: Toward Clarity and Grace, University of Chicago Press.

2. G. W. Hitt, R. G. T. Zegers, S. M. Austin, D. Bazin, A. Gade, D. Galaviz, C. J. Guess, M. Horoi, M. E. Howard, W. D. M. Rae, Y. Shimbara, E. E. Smith, and C. Tur, "Gamow-Teller Transitions to Cu Measured with the Zn(*t*,He) Reaction," (2009).

Example 11.2

An Argument: We conclude that the increase of the diurnal temperature range [DTR] over the United States during the three-day grounding period of 11–14 September 2001 cannot be attributed to the absence of contrails. While missing contrails may have affected the DTR, their impact is probably too small to detect with a statistical significance. The variations in high cloud cover, including contrails and contrail-induced cirrus clouds, contribute weakly to the changes in the diurnal temperature range, which is governed primarily by lower altitude clouds, winds, and humidity.[3]

An alternative form of point-first paragraph is to use a more extended LD structure in which the lead takes several sentences. An example of this is the first paragraph in example 10.2 about nitrogen stoichiometry, in which the lead takes three sentences. The first makes the general argument that organisms have characteristic element ratios, an argument that is sharpened and narrowed to the final clause of the third sentence, which states that these ratios may "predict element mineralization or immobilization during decomposition." That ends the lead with a point the rest of the paragraph expands on.

Another example of an LD paragraph is example 11.3, which comes from a paper about synthesizing complexes between metals and aromatic carbon-60 (C60) structures. This came at the end of the paper's opening, which argued that such complexes are important in nature and that new techniques were becoming available to synthesize them. It makes a critical step in the funnel, narrowing down to the specific problem of synthesizing metal–corannulene ion—molecule complexes.

Example 11.3

Metal-PAH [polycyclic aromatic hydrocarbon] complexes are important as models for surface science and catalysis; PAHs may be used to model finite sections of a carbon surface. There is also evidence that metal-PAH complexes may be constituents of interstellar gas clouds; they have been implicated in the depletion of atomic metal and silicon in the ISM and as contributors to the DIBs [diffuse interstellar bands] and UIB [unidentified infrared bands]. *Increasing interest in metal-PAH systems has thus motivated many groups to produce these species in laboratory experiments.* Dunbar and co-workers were the first to observe metal-PAH ion complexes in gasphase experiments using FT-ICR mass spectrometry. From these experiments, they determined the binding kinetics of a variety of metal and nonmetal cations with PAHs. Our group has produced a variety of metal and multimetal-PAH sandwich complexes using laser vaporization of film-coated metal samples in a molecular beam cluster source. Competitive binding and photodissociation experiments were successful in determining structural information and

3. G. P. Yang Hong, P. Minnis, Y. X. Hu, and G. North, "Do Contrails Significantly Reduce Daily Temperature Range?" *Geophysical Research Letters* 35 (2008): L23815.

relative bonding strengths of metals with benzene, C60, and coronene. In other experiments, we used a laser desorption time-of-flight mass spectrometer to produce a variety of metal oxide and halide PAH complexes as well as mixed-ligand complexes. Experimental work has stimulated new theoretical studies investigating metal binding sites and bond energies on PAHs. Dunbar, Klippenstein and co-workers and Jena and co-workers have been active in this area.[4]

I've italicized the sentence that makes the point of this paragraph; it closes the lead and opens the discussion of research groups that have made metal-PAH complexes. The first sentences discussed why they are important, and thus why researchers would want to synthesize them. The following sentences lay out the history of synthesis efforts. Because this is a long paragraph, the authors used several sentences to build to the point, rather than a single topic sentence.

11.2. POINT-LAST PARAGRAPHS

In a point-first paragraph, you make an argument and then flesh it out. Sometimes, however, you need to assemble an argument, pulling threads together to weave them into a conclusion, producing a point-last structure. These may be either LDR or OCAR.

An LDR paragraph opens with an argument and then develops it, similar to an LD paragraph, but then it wraps up with a synthesis: it's strong at both opening and resolution. A good example is the second paragraph of example 10.2 about nitrogen stoichiometry. That paragraph starts with the strong statement that animals and microbes live in a C-rich, N-poor world; it ends with a conclusion as to what that means—N recycling lags behind decomposition—hence LDR.

An additional example of an LDR paragraph is the following (example 11.4), which is about the treatment of post-traumatic stress disorder (PTSD) following battlefield injuries. The authors suggest that using morphine to ease immediate pain might also help reduce later PTSD, because its anti-anxiety effects can prevent the bad memories from consolidating.

Example 11.4
Although much of the research in the field of pharmacotherapy for the secondary prevention of PTSD after trauma is speculative, there is theoretical evidence that early use of anti-anxiety agents can be effective. Pitman and Delahanty argued that pharmacotherapeutic interventions for the prevention of PTSD will be most effective if medication regimens are implemented after exposure to traumatic events. Morgan and colleagues and other

4. T. M. Ayers, B. C. Westlake, D. V. Preda, L. T. Scott, and M. A. Duncan, "Laser Plasma Production of Metal-Corannulene Ion-Molecule Complexes," *Organometallics* 24 (2005): 4573–78.

investigators have hypothesized that opiates may interfere with or prevent memory consolidation through a beta-adrenergic mechanism. This theory also lends support to the idea that morphine and other opiates may prove effective in the secondary prevention of PTSD after trauma.[5]

In this paragraph the lead is the general argument that anti-anxiety agents should block PTSD; this point is made in the first sentence. The development is the discussion of the papers by Pitman and Delahanty and by Morgan and colleagues. The resolution is the last sentence, which argues that because morphine is an anti-anxiety agent as well as a potent painkiller, it should limit PTSD.

The other way to craft a point-last paragraph is to use an OCAR structure, in which the opening sentence introduces the issue without framing an argument—it just sets the stage. The last sentence synthesizes the material to make the conclusion. Consider example 11.5.

Example 11.5
If the Great Plains mammoths routinely undertook long-distance migrations, then mammoths at all of the Clovis sites in this study should display similar $^{87}Sr/^{86}Sr$ ratios. However, the Dent mammoths display $^{87}Sr/^{86}Sr$ ratios that are distinct from those of mammoths at Blackwater Draw and Miami, demonstrating that the Dent mammoths belonged to a distinct population. Thus, we conclude that Great Plains mammoths did not routinely migrate between northern Colorado and the southern High Plains, which are separated by about 600 km.[6]

The point of this paragraph is that mammoths did not migrate long distances, which is presented in the closing sentence—hence, point-last. The first sentence poses the question (did they migrate long distances?) and the approach to answering it (Sr isotope ratios). It serves as both opening and challenge, but it doesn't answer the question and so doesn't resolve. It acts as a guide to the story but as a classic OCAR opening, rather than as an LDR lead.

Another excellent example of a point-last OCAR paragraph is example 10.1 about population density and watershed N-export. That paragraph opens by arguing that these might be related. The second sentence poses the challenge, asking whether differences in population density could explain the differences in N-export between north and south. Several more sentences develop the action, leading to the resolution: population density cannot explain the patterns of N-export.

5. T. L. Holbrook, M. R. Galarneau, J. L. Dye, K. Quinn, and A. L. Dougherty, "Morphine Use after Combat Injury in Iraq and Post-Traumatic Stress Disorder," *New England Journal of Medicine* 362 (2010): 110–17.

6. K. A. Hoppe, "Late Pleistocene Mammoth Herd Structure, Migration Patterns, and Clovis Hunting Strategies Inferred from Isotopic Analyses of Multiple Death Assemblages," *Paleobiology* 30 (2004): 129–45.

Point-last paragraphs are not terribly common; they might account for 25–30 percent of a paper. Writing is dominated by point-first paragraphs, particularly by TS-D, which is the bread-and-butter paragraph. The complex structures, however, often appear at critical story points—openings, resolutions, and transitions—so you must learn when and how to use them. Additionally, although short paragraphs are usually TS-D, long paragraphs benefit from a resolution to tie them together and remind the reader of the point; they lean toward LDR or OCAR.

I've presented these structures as distinct, but they are not; rather, they form a spectrum from paragraphs with all the power in the first sentence to those with it all in the last—pure TS-D to pure OCAR. Some paragraphs may be hard to classify definitively as TS-D, LD, LDR, or OCAR. Slight shifts in the weighting of a sentence, or of a reader's interpretation, might change how they would define the structure. It's better when the structure is apparent, because if it is unclear, then the point may be, too. It's okay to write point-first paragraphs, and it's okay to write point-last paragraphs, but don't write point-nowhere paragraphs.

11.3. BAD PARAGRAPHS

Because paragraphs tell stories, they can fail for the same reasons that whole stories do. Paragraphs that lack a coherent structure can seem confusing and pointless, as did example 10.3, which illustrated directionless, "stream of consciousness" writing. We fixed that paragraph by breaking it into smaller ones, each of which had a single point; the first paragraph became OCAR and the second TS-D. Anytime you come across a paragraph that seems too long, too rambling, or too incoherent, you need to look for the story arcs and elements, and restructure or break up the paragraph up to highlight them.

As an example, here is a paragraph that is about as bad as it is possible to write (example 11.6). It is about restoring damaged grasslands where phosphorus availability limits plant growth. The researchers did two things: (1) they added compost as fertilizer, and (2) they inoculated a native grass with symbiotic fungi (known as mycorrhizae) to determine whether these treatments would overcome the P-limitation and allow plants to grow well.

Example 11.6:
Adding compost to soil promotes microbial growth, which then increases microbial production of phosphatase enzymes that release plant-available P from organic matter. *Bromus carinatus* is a native grass that can be used in reestablishing California grasslands. Its success in P-poor systems can be stimulated by inoculation with mycorrhizal fungi. However, the effects of mycorrhizal inoculation of *B. carinatus* on P uptake have not been assessed.

Not only is the point of this paragraph completely opaque, so is its structure. Is it point-first or point-last? Does it even have a point? There are several threads of argument that seem to weave aimlessly through it. First, P availability is critical to growing plants and restoring degraded California ecosystems. Second, there are two approaches that may enhance P-uptake by plants: adding compost and inoculating plants with mycorrhizae. Third, *B. carinatus* can be used to reestablish grasslands. Finally, the effects of mycorrhizal inoculation of *B. carinatus* are unknown. Which of these is the point of the paragraph?

The paper intended to connect the first two ideas; the story was about integrating multiple approaches to solve a problem, in this case focusing on increasing P-supply to plants to get them to grow. *B. carinatus* is incidental to that story; it just happens to be a useful species to use. Unfortunately, the authors introduced that point in a way that derailed the paragraph. To the authors, how this all fit together was probably obvious, but it wasn't to a reader. This kind of incoherence frequently results from the curse of knowledge. When authors know too much but write too little, ideas get overcondensed and jumbled.

How do you fix a paragraph like this? The first step is to identify the real story: there are two approaches to restoring degraded grasslands. The second step is to decide whether this needs a point-first or a point-last structure. I argue for a point-first, LD structure. The third step is to pull apart the different threads of this story to clarify their relationships.

Example 11.7

Restoring degraded California grasslands requires adequate supplies of P to support plant growth. Two management approaches have been proposed to achieve this: adding compost and inoculating grasses with mycorrhizal fungi. However, which approach is more effective in enhancing P-uptake and restoration is unclear, and they may work synergistically. Compost not only adds organic P to the soil; it also stimulates microbes that produce phosphatase enzymes and so increases P-release. P uptake by plants, on the other hand, may be stimulated by inoculating them with mycorrhizal fungi.

Bromus carinatus is a native grass that may be useful in restoring degraded California grasslands as it grows extensively throughout the state and tolerates harsh conditions. If it can be supplied with adequate P, it establishes well and starts the restoration process. A question, however, is which approach to enhancing P supply will work best with it, or whether both are necessary.

Instead of one dense, cryptic paragraph, I've broken this into two, added explanation, and ensured that each has a clear point and a clear structure: LD and LDR, respectively. Each forms a coherent story within itself, and together they define a direction for the paper. I argue that as a rule, short is better than long (see chapter 16), but you should take the space necessary to frame critical ideas. If you confuse readers, you lose them. If you need to condense, condense elsewhere.

The key to writing good paragraphs, and fixing bad ones, is the same as for other writing problems. Identify (1) who the story is about, (2) your point, and (3) where you should make it. Put the critical pieces of information in the right places, and use the rest of the text to tie them together smoothly.

EXERCISES

11.1. Analyze published papers

Go to the papers you have been analyzing. Choose a selection of paragraphs in the papers; pick the critical points of opening, challenge, resolution, and transitions, as well as a random sampling of body paragraphs. Define their structures: point-first versus point-last. If point-first, are they simple TS-D or more developed LD? If point-last, are they LDR or OCAR? Evaluate where they use each structure—is there a pattern of where the authors use each type of paragraph? Can you determine why they use a particular structure in each case?

11.2. Write a short article

Go back to your short article and analyze the structure of every paragraph: TS-D, LD, LDR, or OCAR? Do they seem appropriate for the particular location and function they serve? If not (or worse, they have no definable structure), rewrite to give them an appropriate and structure.

11.3. Edit

A. Rewrite Example 11.2 about jet contrails and diurnal temperature variation as a point-last, OCAR, paragraph. Does it work as well?
B. Rewrite Example 11.5 about mammoths as a TS-D paragraph. Does it work as well?

12

Sentences

A sentence tells a story, just the shortest one possible.

It may seem strange, in a book targeted at high-level scientific writing, to go back to a topic that you probably studied in your first class on writing, possibly when you were six years old. But you can't write strong papers with weak sentences. To polish your writing, you need to go back to basics. That means sentences.

Because a sentence tells a story, basic principles of story structure apply. Readers need to meet the characters (opening), learn what they did (action), and what the outcome was (resolution). However, there are grammatical rules that we must integrate with OCAR principles to write good sentences.

Grammatically, a sentence has a subject, verb, and object. To transform a sentence into a story, however, you need to see those *grammatical* units as *story* units that carry out the OCAR functions.

O	Opening: who is the story about?	= Subject
C/A	Challenge/action: what happened	= Verb
R	Resolution: what was its outcome?	= Object

Good sentences present the OCAR elements in the most convenient order possible, establishing a framework and then placing new information into it, allowing readers to process each piece of information in turn. If you give us information out of order, we have to hold it aside until you provide the essential pieces. Imagine helping someone build a house. First you hand them a board, then the nail to pound into it, and only then the hammer to pound it with. If you reverse that order, they'll hang the hammer on their belt and hold the nail in their teeth while they wait for the board to put in place. It's harder that way.

With longer stories, we can choose where to place the emphasis—either in the opening in LD, or in the resolution in OCAR or ABDCE. Most sentences, however, are short enough that you don't have that flexibility. It's in the nature of English that the last word or phrase in a sentence's main clause carries the strongest emotional weight, so simple sentences invariably follow OCAR.

A sentence's condensed form and constrained structure is why Joseph Williams in *Style: Toward Clarity and Grace*," defines the terms "topic" and "stress" for the critical opening and resolution positions, terms that are more specific in how they reflect their function in a sentence.

12.1. OPENING: THE TOPIC

In any story, the opening identifies the characters and setting. In a long story, there may be a number of elements to introduce—people, places, concepts, and so on. However, a sentence is more limited and should deal with a more limited suite of characters, frequently just one—it has a single *topic*.

Whatever you put at the beginning of a sentence, readers interpret as the topic: who or what the sentence is about. Because the topic presents the context for what is to come, it should be a schema or character that readers are familiar with, either because it is common knowledge or because you introduced it earlier. Then you develop the schema by adding new information. If you break that pattern and put new information at the beginning of a sentence, readers may be confused—you're giving them new information but suggesting it's old.

12.2. RESOLUTION: THE STRESS

Endings are always power positions—last words carry the greatest weight. Because of that emphasis, Williams defines the ending of a sentence as the *stress*. That is why I made "the stress" the last words in the previous sentence—I wanted to emphasize the new term. In a multiclause sentence, the ending of each clause is a minor stress position. Use the power of the stress by putting key words there—the main message and new ideas or terms.

To illustrate the power of the stress position, consider the ending of Winston Churchill's famous "We shall fight them on the beaches" speech. Read it aloud and hear where your voice comes down in emphasis.

Example 12.1

"...and even if, which I do not for a moment believe, this Island or a large part of it were subjugated and starving, then our Empire beyond the seas, armed and guarded by the British Fleet, would carry on the struggle, until, in God's good time, the New World, with all its power and might, steps forth to the rescue and the liberation of the Old.

What words did you find yourself hitting? Was it not the end of each clause? With a big punch on the final words—the sentence's overall stress?

Churchill wrote it this way because he knew the words and ideas he wanted to emphasize, so he put them in stress positions.

12.3. PUTTING TOPIC AND STRESS TOGETHER

Some writers seem to think that if they get the right facts into a sentence, readers will get their point. They are wrong. Shifting information between topic and stress changes how readers interpret it. You must put the right information in the right place if you want readers to get your intended point. Consider the following three sentences; they all contain the same facts, but they tell different stories.

Example 12.2

A. Viruses were not studied in the sea until 1989 yet are its most abundant biological entities.

B. The most abundant biological entities in the sea are viruses, yet they were not studied until 1989.

C. The most abundant biological entities in the sea were not studied until 1989: viruses.

In the first sentence, viruses are "old information" that we're learning something new about. That new story is defined in the stress—viruses are the sea's most abundant biological entities. This expands your understanding of where viruses are important.

Sentences B and C, on the other hand, put "biological entities in the sea" in the topic position. They build on a "sea creature" schema, which is probably about fish, rather than on a "virus" schema. Though these sentences both have sea creatures as their central character, they tell different stories. Sentence B emphasizes when they were first studied—isn't it surprising that it wasn't until 1989? Sentence C emphasizes that the surprising new information is that the most abundant entities are viruses. It tells the opposite story from A, inverting old and new information.

Just putting "viruses," "1989," and "most abundant biological entities in the sea" into a single sentence doesn't create one message—it creates six potential messages (of which I showed three). The difference between these sentences and how you interpret them isn't length or language, but structure—what information goes where? You must choose which story you want to tell and structure your sentences accordingly.

Recognizing how readers respond to information in different parts of the sentence offers a tool for controlling those responses. Let's consider another example, one with information that may not be as easy to assimilate.

Example 12.3
Net mineralization represents the nitrogen available to plants because it reflects the difference between microbial nitrogen release and uptake in soil.

Only a minority of you likely knows what "net mineralization" is or have a schema for it. So when I start a sentence with it, the sentence (and ideas) may be challenging. Starting with more familiar concepts makes the sentence easier.

"The amount of nitrogen available for plants is controlled by net mineralization—the difference between microbial nitrogen release and uptake in soil."

This sentence says effectively the same thing, but it starts with an idea that most people understand—plants need nitrogen. It builds off a widely held schema and educates readers about the role of microbes in controlling plant N. All it took to make the story more tractable was switching the topic. Whereas the first sentence might have seemed opaque, this should seem more transparent.

By shifting the order of the ideas in the sentence, I also buried "net mineralization" in the middle, minimizing the weight on the term itself and letting you slide over it. For people in the field, referring to the term, even in a low-emphasis position, may strengthen the message. If, however, I were writing for people outside the field, who don't hold the net mineralization schema, I could leave the term out and simplify the sentence to:

"The amount of nitrogen available for plants is controlled by the difference between microbial nitrogen release and uptake in soil"

This sentence says exactly the same to most people—perhaps more, as it is shorter and contains no unfamiliar information to distract.

If I wanted to define "net nitrogen mineralization" and create a schema for it that I could build on, I would move it to the sentence's overall stress position:

"The amount of nitrogen available for plants is controlled by the balance between microbial nitrogen uptake and release in soil; we define this balance as net N mineralization."

This emphasizes "net N mineralization" as a new term I want readers to remember. I would write it that way in a textbook. I would not, however, write it that way for a specialist journal—experts wouldn't need the term defined. For them it's an established schema. If I wrote the sentence that way for a paper in

Soil Biology & Biochemistry, readers might interpret the writing as that of a novice for whom this *is* new material.

The weighting of words in a sentence follows a consistent order: the stress carries the greatest emphasis, the topic is next, and the middle carries the least. This pattern is described by Roy Peter Clark, in *Writing Tools*, as the 2–3–1 rule of emphasis. Managing this pattern will help you put the right information in the right places; it will guide you in selecting the appropriate topic and stress to write sentences with both clarity and power.

12.4. SUBJECT–VERB CONNECTION

Managing topic and stress is about managing opening and resolution. Writing strong sentences also requires managing the action. Sentences are highly condensed stories; there is no time for a long, gentle opening. Introduce the key characters and then get right to it. The longer the gap between actor and action, the duller and more confusing a sentence becomes. The verb (the action) should immediately follow the sentence's subject.

As examples of tight subject–verb connection look back to example 12.3 and the first suggested rewrite; each immediately follows the subject with the verb:

"Net mineralization *represents* . . ."
"The amount of nitrogen available for plants *is controlled* . . ."

In the second version, the subject is long (seven words), but the verb (is controlled) immediately follows it.

Example 12.4 offers a case where the authors inserted a phrase into the middle of the sentence, set off with commas, and so disconnected the subject and verb, and as a result made it unnecessarily hard to follow.

Example 12.4
The pooled effect sizes, both with and without adjustment for environmental risk factors, were larger for DNA-based than RNA-based viruses.

The subject is "pooled effect sizes" and the verb is "were," but those are separated by the discussion of adjusting for environmental risk factors. To reconnect subject and verb, move that parenthetical out of the middle of the sentence.

"The pooled effect sizes were larger for DNA-based than RNA-based viruses, regardless of whether environmental risk factors were adjusted for."

12.5. MANAGING REAL SENTENCES

The examples I've used so far are straightforward, with few extra words or clauses to add information or nuance. To express more complex thoughts, we need to be able to write more complex sentences, but these often get out of control. Rather than elaborating the message, we end up hiding it, burying it under clutter. The key to writing complex sentences is holding their structure together. Topic, action, and stress need to be well chosen and well placed.

12.5.1. Pick the Right Topic

When we add words or clauses to the beginning of a sentence, we bury the topic and risk that it will be missed or misconstrued. Tightening the structure means picking the right topic. For example, in the following sentence, what is the real topic?

Example 12.5
It has been predicted that the global average temperature will increase at a rate of 0.2°C/decade.

The OCAR structure here is weak because there are two sets of actors and action: (1) someone predicted, and (2) temperature will rise. Why open with the implied nameless people who did the predicting, when the story is almost certainly about global average temperature? Make that the topic:

Global average temperature has been predicted to increase at a rate of 0.2°C/ decade."

This collapses all the action—both the prediction and the increase—into a single short action section, making the sentence clearer. It has a better internal arc structure.

The following is a case where the author only had one potential actor but still managed to bury the sentence's topic.

Example 12.6
In this study, taking advantage of a well-annotated genome map and effective targeted-mutagenesis techniques, we analyzed the role of Bac17 in pathogenesis by *Candida albicans.*

Here, the authors added a long incidental qualifying clause to the beginning of the sentence. Since we don't yet know what is being qualified, we are likely to skim over that material because we don't have a framework for it. It would be better to move the real topic of this sentence closer to the beginning:

"We analyzed the role of Bac17 in pathogenesis by Candida albicans by taking advantage of a well annotated genome map and effective targeted-mutagenesis techniques."

12.5.2. Unburying the Stress

Sometimes the problem with a sentence is that there are words dangling after the real stress. To strengthen such sentences, you need to either delete those extra words or move them into the middle of the sentence, thereby shifting the important words into the stress position. For example, the following sentence (example 12.7) ends weakly. How would you strengthen it?

Example 12.7
The number of commercial products containing nanomaterials has risen rapidly; in 2006 there were only 212 while in 2009 there were over 1000 products on the market.

What is the point in this sentence? The first clause makes the argument that nanomaterial use is increasing, whereas the second elaborates that by showing us how much it has increased. The point to emphasize, therefore, should be 1000, rather than "products on the market." So cut those trailing words; we've already said "commercial products," so we know what the numbers refer to:

"The number of commercial products containing nanomaterials has risen rapidly; in 2006 there were only 212 while in 2009 there were over 1000."

Example 12.8
Plants can increase their resistance to bacterial pathogens by increasing leaf alkaloid concentrations and by synthesizing tannins to bind to bacterial enzymes within plant tissues.

The dangling words "within plant tissues" are confusing. Does "within" refer to where tannins are synthesized or to where the enzymes are? The message to stress is not where it happens but that tannins bind to bacterial enzymes.

"Plants can increase their resistance to bacterial pathogens by increasing leaf alkaloid concentrations and by synthesizing tannins to bind to bacterial enzymes."

The following suffers from a buried stress, but this one can't be solved by deleting the trailing words; the sentence would make no sense if you did that.

Example 12.9
The data did not support our initial hypothesis, as no clear trend in reaction rate with pH was observed.

So what is the problem here? The second clause is supposed to illustrate and support the initial argument. But the important information is the absence of a trend in reaction rate, so that should be stressed. As written, though, the stress is on "was observed"; the information about reaction rate is buried inside the sentence. Rewrite this to stress the reaction rate response:

> "*The data did not support our initial hypothesis, as there was no clear trend in reaction rate with pH.*"

This strengthens the action by focusing on the trend and not on its observation, and improves subject–verb connection, moving the verb to the front of the clause, "as there was."

Possibly the authors thought it important to imply some uncertainty by emphasizing that they didn't *observe* a trend, rather than to state categorically that there wasn't one. In that case, it would be better to rewrite the sentence this way:

> "*The data did not support our initial hypothesis, as we did not observe a clear trend in reaction rate with pH.*"

This puts the important information in the stress and moves the point about observation into the middle of the sentence, the 3-position in the 2–3–1 rule of emphasis. This leaves the qualification in place without making it the point of the sentence.

Many sentences pose even more complex problems. In some, both the topic and stress are buried. That's probably not uncommon in longer sentences, especially those written by developing writers. Example 12.10 offers an example.

Example 12.10
The qualitative agreement between caribou's preference for feeding on young leaves and the trend for protein to decline with leaf age supports the hypothesis that caribou migration is driven by the patterns of leaf-out and maturation spatially and temporally through their home range, rather than by weather.

This is a long, complex, and confusing sentence, but it's a simple idea: caribou migrate to follow the availability of high-quality, nutritious food. We should be able to state a simple idea simply.

Step 1: Fix the topic. Right now the grammatical subject takes 21 words—everything before "supports"; that's way too long and complex. The point is that caribou prefer young leaves, but that is in the middle of the opening clause and is weakly stated. The authors were suggesting that caribou like young leaves because they are protein-rich, but were trying to leave open the possibility that other factors contribute to that preference. They were trying to be careful, but by trying too hard and doing it badly, they ended up being confusing.

To fix this sentence, let's make the topic caribou's food preference: young leaves that are rich in protein. Let's do it in fewer than 20 words: "Caribou prefer to feed on young, protein-rich leaves . . ." This condenses the opening to a simple, nine-word clause that has a one-word topic immediately followed by a verb.

It even leaves causality open; it says that the leaves caribou like are protein-rich, but doesn't specify that is why they like them. Here, we say more by saying less, using structure instead of words to carry the message. It is easy to write yourself into a corner by over-explaining.

Step 2: Fix the stress. In the original version the stress is "weather." It should be the argument that caribou follow food: ". . . the spatial and temporal patterns of leaf-out and maturation."

Step 3: Finish. Having unburied both topic and stress, any additional information can be packaged into the middle. The complete sentence now reads:

"Caribou prefer to feed on young, protein-rich leaves, supporting the hypothesis that migration is not driven by weather but by the spatial and temporal patterns of leaf-out and maturation."

This sentence is now much clearer, and not just because it's shorter. The topic-action-stress is tighter, giving it a better structure. We got there by following a few simple guidelines:

1. The topic should be short and clear.
2. The main verb should follow it immediately.
3. The key message should come at the stress.

These apply to individual clauses as well as to the sentence as a whole. Remember the hierarchical structure of story arcs. The final sentence of example 12.10 now has three clauses, each of which is tight and clear.

12.6. LONG SENTENCES

Most books on writing argue that short sentences are easier to read than long ones. They aren't wrong, but that is one of the simplifications you should unlearn. The example from Winston Churchill was 141 words long and was neither short nor bad writing—a crime few would dare accuse Churchill of.

Good, clear sentences can be short or long, and the best writers use a mixture of both. The key to writing a good long sentence is holding together the structure, but straight OCAR won't work. If readers have to wait to the end of the sentence to get the point, they will get lost. To write a good long sentence, you need to use an LD structure: make the key point in a short initial main clause, and then add others that add depth and nuance. This is what Roy Peter Clark describes as a "right opening sentence." Even though we lean toward short sentences in science writing, it's a useful skill to be able to craft a clean long one. For example, I fixed the subject–verb connection in example 12.4 by giving it an LD structure—I made the first clause tell the story and added the elaboration in a subordinate clause.

Here's an example of a right-opening sentence from a paper that reconstructed the historical climate of east Africa by analyzing the past level of Lake Tanganyika.

Example 12.11
At the beginning of the second millennium AD, lake level at Lake Tanganyika
fell and remained relatively low during the period from ~1050 – 1250 AD,
which corresponds to the timing of the MWP [Medieval Warm Period] in
many locales, albeit with a later onset than in some areas.[1]

This sentence is long (more than 40 words), with a complex multiclause structure, but the story is simple and clear: Lake Tanganyika was low for several hundred years after 1000 A.D. The opening clause sets the stage for the main message,
"Lake Tanganyika fell." *Fell* is a powerful verb that immediately follows the subject. From there, the authors added additional clauses that created extra dimensions. The sentence uses a right-opening LD design. It sketches in the main story
and then colors in the picture, essentially building and then growing a schema.
Although long, this sentence feels simple because its structure works.

   Consider what this sentence would look like if we wrote it to put the key action
(the lake fell) in the stress of the overall sentence:

*"During the early part of the second millennium AD, from ~1050 – 1250 AD, a
period corresponding to the timing of the MWP in many locales, albeit with a
later onset than in some, lake level at Lake Tanganyika fell and remained relatively low."*

   This is much harder. As readers work their way through this sentence, they are
asking themselves "Who's this story about?" or "What's the point?" They would
have no old-information framework on which to hang the information about
timing, and so they would lose the thread. Get to the topic quickly, then to the
action, and then add nuance if needed.

   Here's another example of a long sentence that uses the right-opening approach
to tell a story about two GTPase proteins.

Example 12.12
We focused on two members of this family: Rab5, which controls transport
from the plasma membrane to the early endosome and regulates the
dynamics of early endosome fusion, and Rab7, which governs membrane
flux into and out of late endosomes.[2]

1. S. R. Alin and A. S. Cohen, "Lake-Level History of Lake Tanganyika, East Africa, for the
Past 2500 Years Based on Ostracode-Inferred Water-Depth Reconstruction," *Palaeogeography,
Palaeoclimatology, Palaeoecology* 199 (2003): 31–49.

2. M. C. Pascale, S. Franceschelli, O. Moltedo, F. Belleudi, M. R. Torrisi, C. Bucci, S. La Fontaine,
J. F. B. Mercer, and A. Leone, "Endosomal Trafficking of the Menkes Copper ATPase ATP7A
Is Mediated by Vesicles Containing the Rab7 and Rab5 GTPase Proteins," *Experimental Cell
Research* 291 (2003): 377–85.

This sentence reads easily because the authors structured it carefully. Their first two words are a short subject and an action verb: "We focused." Then they hit us with critical information in the stress of the opening clause; they focused on "two members of this family." That clause fully framed the story and the colon suggests that the authors will follow up by describing each member, which they do. First they name Rab5 and then describe it, and then they come back to Rab7 and describe it. This created a sentence with a doubly branched, right-opening structure. That sounds (but doesn't feel) complicated; it worked because the authors held the structure together when it easily could have fallen apart.

As you can see from the examples, the secret to writing strong sentences is the same as for writing strong papers and paragraphs: make the OCAR elements clear and put them in their right places. Develop story elements into coherent arcs placed in series. Find the topic, make it the subject, and move it toward the beginning of the sentence. Find the action verb and connect it closely to the subject. Find the stress and move it to the end of the main clause. If you have additional material to add, move it to the right so that it modifies, rather than intrudes in, the main story. By doing so, you should be able to take even painfully confusing and complex sentences and make them tight, clear, and easy to read.

EXERCISES

12.1. Evaluate published papers

Pick sentences from several of the paragraphs you analyzed for the exercises in chapter 11. Evaluate what the authors chose to put at the topic and stress positions. Were they good choices? If the authors used more complex sentence structures, did they maintain strong subject–verb connection? If there are sentences that are poorly structured, can you rewrite them to make them clearer?

12.2. Write a short article

Go through every sentence in your article and examine sentence structure. Have you chosen the right topic and stress? Have you ensured tight subject–verb connection? Rewrite them, if necessary, to strengthen them.

12.3. Write new sentences

A. I argued that taking three elements allows you to create six possible stories. Take the elements I used in example 12.2 (viruses, 1989, and biological entities in the sea) and write the three that I did not include.
B. Take the following pieces of information to write three different sentences with three different stories: benzene, cancer, groundwater.

12.4. Revise the following sentences to strengthen them

A. Due to uncertainties resulting from interferences in the X-ray microanalysis, it remains unclear what the crystalline nature of kryptonite is.

B. By reducing diffusion and increasing physiological stress, drought reduces soil microbial activity and causes a build-up of biodegradable C that is rapidly respired upon rewetting.

13

Flow

Good writing has an aliveness that keeps the reader reading from one
paragraph to the next.

—WILLIAM ZINSSER, *On Writing Well*

Best-selling novels are often described as "page turners." Best-cited papers
and best-funded proposals are the same. They draw readers in and lead
them through the story—they flow. A break in that flow can derail a reader and
abruptly change a piece from "page turner" to "re-turner" with a rejection letter
attached.

There are two approaches to creating flow. The first is to write paragraphs
where all the sentences are on the same team—dealing with a coherent theme and
working together for a common goal. The second is to write sentences so the team
forms a relay—each passes a baton at the transition, allowing an idea to flow
cleanly from start to finish.

When you write a paragraph, the opening sets its theme. As long as every sentence has a topic that fits into that theme, the paragraph will hang together. All
paragraphs need this kind of thematic coherence, and when they lack it, we suffer.
For example, the stream-of-consciousness paragraph about California fisheries in
example 10.3 had eight separate topics that didn't fit a pattern; it was incoherent.

In contrast, in this paragraph the topic of every sentence is either "paragraph" or "sentence," ideas that fit comfortably together; I think it is coherent.

The greater challenge, which I focus on in this chapter, is helping readers avoid derailing at transitions—helping them follow you through your arguments and between the story arcs. If you lose readers at those critical points, you may lose them for good. To carry readers through, sentences need to link seamlessly to each other. Achieving this flow involves a delicate balancing act. Each sentence must tell a coherent story, but each must also function within a paragraph, advancing the larger story. This requires tying together stress and topic, weaving old and new information into an unbroken chain.

The critical element in building this chain is managing the topics. These should refer to information that is familiar from earlier text. This is a backward-looking function. But the topic also tells you what the story is going to be about, which is a forward-looking function. For a story in a series, a sentence in a paragraph, the topic is like the Roman god Janus (figure 13.1), simultaneously looking backward and forward, connecting past to future.

When people read a story, they need the opening of each arc to provide that Janus function—both context and launching point. When your topics achieve that, your ideas will be easy to follow. When they don't, they may feel disjointed.

As a starting example, consider the following sets of sentences.

Example 13.1
Molecules are comprised of covalently bonded atoms. Molecules' reactions are controlled by the strength of the bonds. Molecules, however, sometimes react slower than bond strength would predict.

Figure 13.1. Janus, the god of doors and beginnings.

Example 13.2
Molecules are comprised of covalently bonded atoms. Bond strength controls a molecule's reactions. Sometimes however, those reactions are slower than bond strength would predict.

These short paragraphs say the same thing. You probably felt, however, that 13.1 is choppy and reads like a list of facts, whereas 13.2 was smoother and makes a coherent story. Why are they different? How do they vary structurally to produce those responses? Let me sketch the sentences out and identify their topics and stresses.

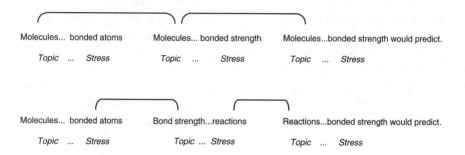

In example 13.1, the topic of each sentence is "molecules." The sentences form a list of facts about molecules without any structure to those facts. In example 13.2, the topic of each sentence ties to the stress of the previous one. That links past to future, fulfilling the Janus function and creating flow. Linking topic-to-topic creates a list of statements about a thing, while linking stress-to-topic creates a story about that thing (figure 13.2).

Pure lists are hard; the reader has to figure out why the facts are important and how they fit together. To make several points about a single topic, we need an LD story structure, which gives an overview to define the theme for the list to follow. Then several sentences can link back to the theme developed by the lead.

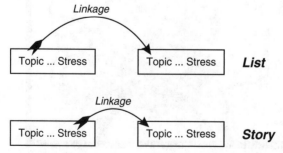

Figure 13.2. Creating a list versus a story: the role of topic and stress.

A story makes the reader's job easy, highlighting each fact and its relationship to others. It builds a framework for each piece of information, socketing new into old. To create these links, you may end and begin sentences with a single word, as I did between the second and third sentences in example 13.2, repeating "reactions." More often, you will repeat an idea or a theme, as I did between the first and second sentences; I picked up the idea of "bonds" but transformed "bonded atoms" into "bond strength." Those are clearly on the same team but not quite the same idea. The two faces of Janus are not identical.

Let's explore how to build flow by looking at real examples, rather than the simplistic ones about molecules. Example 13.3 is a passage in which the individual sentences are reasonably well written but where there is a flow problem.

Example 13.3
Salvage logging is an increasingly common way of harvesting forests that have been attacked by insect pests. In salvage logging, trees that have been attacked are selectively harvested. Cavities in standing dead trees serve as nesting sites for birds. The population biology of cavity-nesting birds is therefore likely affected by salvage logging.

Did you feel a break in the flow? The first sentence opens with a clear topic. salvage logging. But then, the third sentence introduces a new topic and a new story arc—dead trees and cavity-nesting birds. Those two story arcs don't connect: To make this passage flow, we need to link the arcs together. How? The easiest way is to reach back and grab the idea of "harvesting" from the first arc and make it the topic of the second. "The dead trees that are harvested, however, can provide cavities that are nesting sites for birds." Now read the whole section:

"Salvage logging is an increasingly common way of harvesting forests that have been attacked by insect pests. In salvage logging, trees that have been attacked are selectively harvested. The dead trees that are harvested, however, can provide cavities that are nesting sites for birds. The population biology of cavity-nesting birds is therefore likely affected by salvage logging."

The flow of ideas is now seamless and the structure of the argument easier to follow. As with example 13.2 about molecules, I didn't repeat an exact word; rather, I grabbed an idea—trees that have been harvested. Note that I made the passage flow by making the subject of the transitional sentence longer, normally a bad thing. But by breaking the rule, I fulfilled the principle of making the reader's job easy.

To finish analyzing this passage, consider it as a story. This is a clear OCAR paragraph. The first sentence provides the opening and topic—it's going to be about salvage logging and its effects. There is action that fills you in on these effects. Then, there is a resolution that makes the paragraph's point and frames the challenge for the paper—salvage logging affects bird population biology. This paragraph forms a tight circle by opening and resolving with the same words, "salvage logging," but it pulls that into a spiral: by the end, your understanding of salvage logging has changed.

The second issue to look at in this paragraph is when it does and does not explicitly link sentences. Look at the first two sentences:

Salvage logging is an increasingly common way of harvesting forests.
In salvage logging, trees that have been attacked are selectively harvested.

Both open with "salvage logging," and so form a list, rather than a story. Is this a problem? Not in this case. These two sentences work together as a unit. The first makes a general statement that the second elaborates. By using the same topic for each, this arc gains thematic coherence; a short list is fine. But when the paragraph shifts arcs, it needs to help readers across the transition.

Example 13.4 raises different issues in developing flow.

Example 13.4
Groundwater level is an important control of the fate of contaminants within the groundwater: are they taken up plants and microorganisms in the surface soil? Industrial landscapes are frequently disturbed, and the effects of this disturbance on the system's ability to process contaminants has not often been studied. Low water tables could reduce an ecosystem's ability to process groundwater contaminants by moving these contaminants out of the reach of plant roots and microorganisms in the surface soil.

Does this paragraph flow? No. The first two sentences don't connect thematically, and their stress-topic linkage is nonexistent: "Groundwater . . . microorganisms in the surface soil?   Industrial landscapes . . ."

So how do we improve the flow between them? This is a trick question—we don't. The real linkage here is between the first and third sentences. "Groundwater . . . microorganisms in the surface soil?   Low water tables . . ." These sentences connect because there is coherence among the terms *groundwater, surface soil,* and *low water tables*: they all relate to level in the soil column. So make the connection by *deleting* the second sentence:

"Groundwater level is an important control of the fate of contaminants within the groundwater: are they taken up plants or microorganisms in the surface soil? Low water tables could reduce an ecosystems ability to process groundwater contaminants by moving these contaminants out of the reach of plant roots and microorganisms in the surface soil."

Now there is no break in the flow. The first sentence poses a question about groundwater level, while the next picks that up and discusses possible answers. These sentences thematically link stress to topic.

The original second sentence about industrial landscapes may make an important point, but if so, put it somewhere else. Make it its own story arc and tie it back into this one. Don't just squash it in here to introduce the characters of "industrial development" and "landscape disturbance." Find a way to give them a proper introduction in their own space.

This example illustrates how you sometimes have to take a global view of a problem to ensure you have accurately diagnosed it. In this case, it was mixed-up, rather than poorly connected story arcs. Diagnosing and solving the *real* problem solved the *apparent* problem. Fixing the arc structure fixed the flow.

If I had taken a narrow approach, I could have forced connection by rewriting the second sentence in the passive voice to reach back and grab the first sentence's stress.

> "*Groundwater level is an important control of the fate of contaminants within the groundwater: are they taken up plants or microorganisms in the surface soil?* These organisms may be disturbed by industrial development, but such effects on contaminant fate have not often been studied. *Low water tables could reduce an ecosystems ability to process groundwater contaminants by moving these contaminants out of the reach of plant roots and microorganisms in the surface soil.*"

This makes the first sentence's stress into the second one's topic, and moves "industrial development" to the right into new material space.

The first two sentences now connect, but the second and third do not. I haven't solved the connection problem, I've just pushed it down the paragraph. That also pushes the story into a different direction (probably the wrong one). This used a tactical approach to solve a problem better addressed by strategy—restructuring the flow of my thoughts. Revising should be a top-down process that never loses sight of the big picture even while you are dealing with details.

13.1. LINKING PARAGRAPHS WITHIN A SECTION

A paragraph break indicates that you are shifting ideas and moving into a new story arc. But readers expect the new paragraph to build off the previous one, developing a larger story. To make them connect, you need to use the same stress-topic, resolution-opening strategies we discussed for individual sentences. Example 13.5 illustrates such an approach from a paper on the fate of nanomaterials in the environment.

Example 13.5
Most previous studies on the aggregation of nanomaterials have focused on the behavior of uncoated particles in dilute aqueous suspension with few contaminants present. However, particles released into natural settings are likely to have organic coatings and will be exposed to humic acids in solution. The interactions of coatings and humics can alter surface redox potentials and particle aggregation.

Nanotitanium oxide particles coated with hydrophobic coatings showed higher aggregation in a medium with high concentrations of humic acids than in control waters. Particles coated with hydrophilic coatings, however, showed reduced aggregation in the presence of humics.

Do these paragraphs flow? I don't think so. Yet clearly there is a connection— the first paragraph lays out general arguments about particle behavior in nature, while the next picks up on that and lays out the results of this research. The problem is the lack of overt connection.

The first paragraph focuses on the aggregation of nanomaterials. It uses an OCAR structure that resolves with an argument about coatings, humics, and aggregation. The topic of the second paragraph, however, is specific: nanotitanium oxide particles. To make this work as a story, we need to tie the opening of the second paragraph back to the resolution of the first. We can do that by grabbing the theme of "aggregation" and restructuring the second paragraph to emphasize it.

The interactions of coatings and humics can alter surface redox potentials and particle aggregation.

Aggregation of nanotitanium oxide particles coated with hydrophobic coatings was higher in media with a high concentration of humic acids than in control waters. . . .

This now explicitly links the paragraphs by using the same word to open the second paragraph that resolved the first.

The following example suffers from the same problem and can be solved using the same general approach, but it illustrates that linking paragraphs isn't always as easy as grabbing a single word. Sometimes you need to go with a larger concept.

Example 13.6

Any trait that increases a bacterium's ability to survive an environmental stress, such as heavy metals or antibiotics, can be considered a stress-adaptation mechanism. Traditionally, however, studies have focused on internal mechanisms of adaptation: either a bacterium's ability to either repair cell damage (e.g., DNA repair) following stress or on mechanisms that make the cell more able to resist the damage in the first place (e.g., producing chaperones and transporters). However bacteria also have mechanisms that work outside the cell to reduce the intensity of stresses in the first place.

E. coli provides an excellent model system for studying how the relative physiological costs of different stress-adaptation mechanisms vary between heavy metal and antibiotic stressors. We understand *E. coli*'s metabolic pathways well enough to assess the full energetic costs of internal adaptation mechanisms such as DNA repair and exporting toxic agents from the cell vs. external mechanisms such as producing chelating agents to bind heavy metals and extra-cellular oxidase enzymes to break down antibiotics before they enter the cell.

These paragraphs don't feel connected. The first paragraph discusses mechanisms of microbial response to environmental stress. The second tells a story about *E. coli* and its utility for studying the physiological costs of stress adaptation. These are related to each other, but as you start reading the second paragraph, it doesn't help you see the link. There is no connection between the resolution of the first paragraph and the opening of the second. What connects them is that they share the same stress: responses to environmental stressors. If topic-topic makes a list, and stress-topic makes a story, what does stress-stress make? Confused readers.

To make the writing flow, we need to connect stress and topic. Once again, the easiest approach is to open paragraph 2 by repeating the stress from the previous. That is the idea that bacteria have multiple mechanisms for dealing with environmental stressors. I would write the opening sentence of paragraph 2 as:

"An excellent model system for studying the relative costs of different stress-adaptation responses to heavy metals and antibiotics is E. coli."

This gets the ideas from the first paragraph (stress adaptation) and provides context for driving the story forward. It moves the new information—*E. coli* as a model system—to the opening sentence's stress, from where it flows smoothly into the next. This version uses a long subject—19 words—and so breaks the short-subject rule I discussed in chapter 12, but this is a case where following the principle of making ideas connect means breaking the rule.

When you are trying to make paragraphs connect, don't forget example 13.4 on groundwater. Sometimes the simple "reach back and grab" approach creates local flow but global disruption. The problem may not be that the second paragraph starts in the wrong place but that the first points in the wrong direction. In these cases, fix the first paragraph.

For example, if we wanted the second paragraph in example 13.5 to be about *E. coli* as a model system, we could rewrite the end of the first paragraph:

"However bacteria, such as E. coli, also have mechanisms that work outside the cell to reduce the intensity of environmental stressor in the first place.

E. coli is an excellent system for studying the relative costs of these external mechanisms relative to internal mechanisms in response to heavy metal and antibiotic stressors. We understand E. coli's metabolic pathways . . ."

In this version, I introduce *E. coli* as a model in the first paragraph, and then use the whole first sentence in the next to reach back and grab. I can't tell you which of these approaches is best—only you know your story.

In many cases, as long as sentences fit into the theme set by a lead or topic sentence, the writing will feel coherent. But to carry readers over the rough spots, you need to do more. Make your ideas connect so tightly that we can't get lost. By linking stress and topic, resolution and opening, you can tie together sentences and paragraphs and make the sweep of your arguments compelling. You can make your papers and proposals page turners.

EXERCISES

13.1. Evaluate published papers

Pick two of the papers you've been evaluating: one that reads well and smoothly and one that doesn't. Evaluate how each manages the flow of ideas between sentences and between paragraphs. Do they use the techniques discussed in this chapter to develop flow? If they don't are there places where you could rewrite them to enhance the flow and ease the writing?

13.2. Write a short article

Revisit your short article and look to see whether you could enhance the flow of ideas by working more actively on linking your sentences and paragraphs together. If you can, do it.

13.3. Edit

A. Revise these sentences so the ideas flow smoothly.

Studies comparing iron-resistant and sensitive cell lines confirmed that protein X17 is denatured in the presence of Fe. Protein X17, however, reverts to its native form when cellular Fe concentrations decrease.

B. Revise the following paragraphs so that the ideas link clearly.

The TREE2 promoter is upstream of the *tryb* gene. However, Silva et al. (2008) showed that the STEM3 promoter is also upstream of this gene and appeared to be involved in *tryb* regulation. We showed that both are necessary and they interact to further upregulate *tryb*: knocking out TREE2 reduced transcription by 50%, knocking out STEM3 reduced it by 75%, and knocking out both prevented transcription entirely.

When LEA was absent, however, even with both promoters intact, transcription rates were only 50% of control levels. It appears that LEA interacts with TREE2 and STEM3 to induce greater promotion than either can effect on their own.

14

Energizing Writing

Don't say the old lady screamed. Bring her on and let her scream.

—MARK TWAIN

Good stories are driven by action. But the drive comes from seeing and feeling it; the old lady has to come on stage and scream—no disembodied howls in the distance. Writers condense that idea to the mantra of "show, don't tell." In science writing, the two C's of SUCCES—credible and concrete—both emerge from showing. We *show* the reader our data, and we *show* them our logic. Isn't the phrase "data not shown" always a little suspect?

Within a sentence, showing action is the job of verbs and it's an important job. Good writers use their verbs well, imbuing their papers with life. Bad writers use them poorly, stealing energy from the story, leaving it dull and listless. While bureaucrats are the grand masters of turgid text, some scientists compete with them for the title. There are many ways to overburden your writing, including three notable ways to emasculate your verbs: (1) passive voice, (2) fuzzy verbs, and (3) nominalizations.

14.1. ACTIVE VERSUS PASSIVE VOICE

The simplest structure for any story is straight OCAR, which gives the reader information in the sequence that they can most easily process: who did it (O), what they did (A), and what happened (R). The sentence structure that most directly matches OCAR is as follows.

John	called	Jane
Actor	Action	Acted-on
Subject	Verb	Object

This is the active voice. It is clear, concise, and direct. It is also visual and evocative. You can see the actors because they are named up front, and you can visualize the action because it is carried in a verb that immediately follows the subject. Hence, Strunk and White's commandment: "Use the active voice."

Sometimes, though, we don't want to tell a story about the actor but about the acted-on; we want to talk about Jane, not John.

Jane	was called	by John
Acted-on	Action	Actor
Subject	Verb	Object

This is the passive voice. To create it, we make the acted-on the sentence's subject and make the verb by coupling some form of "to be" to the action verb; in this case, "was called." The passive voice is a powerful tool. It allows you to control who or what, in a sentence, the story is about. It allows you to select the grammatical subject and object of the *sentence* relative to the actor and acted-on of the *story*. It allows you to control what goes in the topic and what goes in the stress. And it does all that without changing the action. Whether you write "John called Jane" or "Jane was called by John," John still made the call. Only our perspective has changed.

But the passive voice carries a price: it weakens the story structure. "With an active verb, the subject of the sentence is doing something. With a passive verb, something is being done to the subject of the sentence. The subject is just letting it happen" (Stephen King, "On Writing").

Without an actor front and center, action is intangible. All we can do is say the old lady screamed; in the passive, we leave her off stage.

14.1.1. Controlling Perspective

Because the passive voice is weaker storytelling than the active, we should avoid it as a matter of course, but it has several good uses. The first is in controlling perspective: who the sentence is about. Being able to shift a sentence's topic between actor and acted-on is critical for developing effective story arcs and flow, as I discussed in chapter 13 and illustrated in example 13.3. There I rewrote the transitional sentence to start with a passive expression:

> "In salvage logging, trees that have been attacked are selectively harvested. The dead trees that are harvested, *however, can provide cavities that are nesting sites for birds.*"

That created a single entity that was the subject for a larger, active voice sentence. Often though, you may need to write the entire sentence in the passive voice to get the appropriate subject up front, as illustrated by the following sentence pairs.

Example 14.1
    Active: A magnetospheric source produces variable electric fields.
    Passive: Variable electric fields *are produced* by a magnetospheric source.

Example 14.2
    Active: Soil porosity influences water retention.
    Passive: Water retention in soil *is influenced* by porosity.

In each case, if this were the entire story, the active voice version would be better—a stronger message and a shorter sentence. But if we are telling a story about variable electric fields or water retention in soil, the passive would put the right term in the right place and so would be the voice to choose. In allowing you to shift perspective this way, the passive voice shines. It's a tool that weakens a single sentence, but in a way that can allow it to fit more snugly into a paragraph, strengthening the whole.

14.1.2. Hiding the Actor

The strength of the active voice is that it forces you to make the actor and action clear. The weakness of the active voice is that it forces you to make the actor and action clear. Sometimes, we don't want to or need to name the actor. The passive voice can do this. The classic example is "mistakes were made," which is used frequently by politicians and bureaucrats to dampen the intensity of the action and dodge blame. Converting "mistakes were made" to the active voice requires putting an actor up front—identifying who screwed up.

Being able to leave the actor off stage, however, is useful for solving a variety of writing problems. For example, even in Materials and Methods sections it has become generally acceptable to use first-person active voice, for example, "We collected samples." That is fine when it's true. However, in a paper, "we" means the authors. What do you say if there were technicians or interns who helped with the work but aren't coauthors? I know some projects where dozens of people contributed to sample and data collection, and the authors may not even know who specifically collected which samples! Using the passive allows the author to cleanly and honestly tell us what we need to know: samples were collected.

We also use the passive to refer to work when the specific attribution isn't important. For example, I wrote a paper once evaluating when the composition of soil microbial communities affects ecosystem processes such as plant litter decomposition.

Example 14.3
It has been argued that such processes should be insensitive to microbial community composition (Schimel 1995).[1]

I deliberately wrote that using the passive phrase "It has been argued" because I had made that argument in an earlier paper but my thinking on the subject had evolved. The new paper was modifying the argument to reflect that evolution. If I had used the active voice, I would have had to say "In a previous paper, I argued . . . but now I think I was wrong," which I thought sounded bad, even if changing our thinking is what we are supposed to do. It might have trivialized the point by making it seem that I was having a private debate with myself. In fact, I wasn't; other papers had made similar arguments or reinforced those I made in the first paper. Because the passive voice doesn't specify who or how many people made the argument, I could leave that open. Even citing my earlier paper didn't mean it was the only one to make the argument. Using the passive to say "It has been argued" allowed me to avoid these issues.

For those times that we need to say what happened and not who did it, the passive is an effective tool. However, it was a long-standing tradition that scientists should divorce themselves from the work by describing their actions in the passive voice, as illustrated in example 14.4.

Example 14.4
When expression of *Chla* and *Chlb* were compared, similar patterns of transcript abundance were observed in plants at different developmental stages.

Someone did the comparing and observing, and most likely it was the authors. So why not make this clearer by naming the actor and shortening the subject? "When we compared *Chla* and *Chlb* expression, we found similar patterns of transcript abundance in plants at different developmental stages."

1. J. P. Schimel, J. Bennett, and N. Fierer, "Microbial Community Composition and Soil N Cycling: Is There Really a Connection?" In: *Biological Diversity and Function in Soils,* ed. R. D. Bardgett, D. W. Hopkins, and M. B. Usher (Cambridge University Press, 2005), pp. 171–88.

It's possible that this was one of those cases where someone who wasn't an author did the work. More likely, though, the authors were following the passive voice tradition. Why did scholars insist on using the passive? Why ban the more powerful storytelling tool? ? The argument was grounded in concerns of scientific objectivity, as expressed in this excerpt: "Using the passive voice in scientific writing allows the researcher to stand at a distance from his or her work. By standing at a distance, an unbiased viewpoint is much more likely to be reached. An unbiased viewpoint encourages a world view and an open mind, surely prerequisites for good science."[2]

This is impassioned plea, and it contains some important truths—but some equally important fallacies. It argues that writing in the passive forces you to remain at a distance from your data and be dispassionate and objective about your work. I agree that objectivity is a prerequisite for good science. However, objectivity does not come from how you treat your *writing* but from how you treat your *data*. The idea that by removing ourselves visibly from the writing we remove our prejudices and imperfections is plain wrong. We did the work, and we wrote the words. They are inextricably ours. You can't change that by changing the writing voice.

True objectivity grows from Anne Lamott's advice in chapter 2: listen to your characters. Be attentive to your data and allow the story to flow from them. Once you have done that, tell the story in the clearest, most effective language possible.

It is a principle that all tools in English have their value, including the passive voice. Even Strunk and White moderate their dictum about using the active voice: "This rule does not, of course, mean that the writer should entirely discard the passive voice, which is frequently convenient and sometimes necessary."

As with all tools, you must know their strengths and limitations to make good decisions about when to use them. The passive voice is for when you need to make the acted-on the subject of the sentence or when you have an honest reason to avoid naming the actor. Use it for those jobs. Otherwise, listen to Strunk and White: use the active voice.

14.2. FUZZY VERBS

Science writing isn't supposed to use colorful language to evoke image the way fiction does, although it can be more colorful than most of us make it, as illustrated in examples 5.8 and 7.1. We are, however, supposed to be clear, and verbs that show action make writing clear. Verbs that mask the action are weak and can be confusing. Consider the following.

Example 14.5
Controls on the expression of homeobox genes have been evaluated in several model systems.

2. S. R. Leather, "The Case for the Passive Voice," *Nature* 381 (1996): 467.

Here the verb is "evaluated;" it's passive, but that isn't the problem. The problem is that this tells us something happened—controls on expression were *evaluated*—but what we really want to know is either how they were evaluated or how they differ between organisms.

This sentence is the opening of a paragraph that goes on to tell how controls differ develops into an interesting story. But the opening would be stronger if it identified what that story is going to be: varying patterns of control. An opening sentence that uses an action verb to introduce that would be:

> *"Homeobox gene expression is regulated differently among plants, fungi, and animals."*

Not only does this make the model systems concrete, it uses a stronger verb and has a stronger message—"is regulated differently." That makes it obvious that the paragraph will discuss how expression varies across these organisms.

The verb is still passive, but the passive allows us to make this sentence about homeobox genes rather than about plants, fungi, and animals. We could turn it around to activate the verb, but that would make it into: "Plants, fungi, and animals regulate homeobox gene expression differently." This makes the topic the organisms, instead of the genes. It also forces apart the verb and adverb. The original, passive voice version avoided that.

Another example that suffers from a fuzzy verb problem is example 14.6.

Example 14.6
Herbivores facilitate the invasion of exotic grasses by mediating competition between exotic and native plants.

The verbs are "facilitate" and "mediate," but we are likely to ask "what do herbivores *do* to mediate competition?"

> *"Herbivores preferentially eat native plants, giving exotic grasses a competitive advantage that allows them to invade."*

This sentence now uses verbs that show action: "eat," "give," and "invade." It says what the animals physically do; they eat native plants. This allows exotics to invade the gaps created. If this sentence were the opening of a paragraph, it would now effectively introduce the characters (herbivores, native plants, and invading exotic grasses), the actions, and the challenge (how herbivores influence invasion). It even puts the critical action, "invade," in the stress position to emphasize it.

Fuzzy verbs say that something happened but not what; action verbs show you what (see table 14.1). Action verbs are powerful, concrete storytelling tools. They make your writing more interesting, which is nice, but also clearer, which is vital.

14.2.1. Fuzzy Hypotheses

The worst place for a fuzzy verb is in a hypothesis, yet many are wishy-washy and unfalsifiable. I've read proposals with hypotheses like the following.

Table 14.1. FUZZY VERBS VERSUS ACTION VERBS

Fuzzy Verbs (Weak)

Occur	Facilitate	Conduct	Implement
Affect	Perform		

Action Verbs (Strong)

Modify	Increase	React	Accelerate
Accomplish	Decrease	Inhibit	Migrate
Create	Invade	Disrupt	

Example 14.7
Microbial community composition is controlled by the chemical nature of plant inputs, water availability, and soil chemistry.

Here the verb is the passive and fuzzy "is controlled," and this is a truism rather than a falsifiable hypothesis. Is it conceivable that microbial community composition is not controlled by plant inputs, water, and soil chemistry? Fuzzy hypotheses almost guarantee that your proposal will end up on the "do not fund" list. To make a hypothesis compelling, you need to use concrete verbs that make a testable statement. To transform example 14.7, consider an alternative:

"The chemical nature of plant inputs is the single strongest control on the composition of soil microbial communities and on their distribution across the landscape."

This is in the active voice and the verb is simply "is." It is a declarative statement—the chemical nature of plant inputs either is or is not the single strongest control; we can test that. This version doesn't ignore other factors, but it puts them in perspective. This was the actual hypothesis of a proposal, a successful one.

I think people use fuzzy verbs when they are afraid that if they make strong statements, someone may challenge them or they may be wrong. If people feel challenged, you have engaged their interest, and that is good. Challenging proposals sometimes get funded; boring ones never do. Also remember, you are a scientist—it is not your job to be right. It is your job to be thoughtful, careful, and analytical; it is your job to challenge your ideas and to try to falsify your hypotheses; it is your job to be open and honest about the uncertainties in your data and conclusions. But if you are doing cutting-edge work, you are not always going to be right.

You may have some aspects of the system right but others wrong; your piece of the system may be counterbalanced by others; you may even have misinterpreted your data. As long as you did it with honesty, integrity, and intellect, you *did* right, even if you *weren't* right.

People must be able to understand your work and how it influences our understanding of nature. Being concrete and challenging may achieve that and move the field forward, regardless of whether you are right. Being nebulous and timid to avoid being wrong ensures that your work will contribute little. As a result, it will likely be rejected or uncited. One of my mentors, a leader in the field, took gleeful delight in tossing out ideas and stirring up the pot; some ideas were brilliant, others off the wall. He left it to others to figure out which were which. The brilliant ideas stuck and motivated new research; the others faded. Being interesting is ultimately more important than being right.

14.3. NOMINALIZATIONS

Fuzzy verbs are energy thieves. They steal energy from the action by telling, rather than showing. You can, however, go a step further and kill the action entirely. Using a strong verb, you might say something like the following.

Example 14.8
We investigated the effect of elevated CO_2 on plant growth.

Here the action is expressed in a verb, "investigated," but many would write this sentence as: "We conducted an investigation of the effect of elevated CO_2 on plant growth." This sentence has a verb—the fuzzy "conducted." But did you conduct an investigation, a train, or an orchestra? The action is contained in "an investigation," but that is a noun. This sentence names the action and introduces a new verb that hides it.

This process of turning a verb into a noun is known as creating a nominalization. As a result of using a noun rather than a verb to describe action, example 14.8 lost energy and gained length, but contains no more information. That is all bad, yet using nominalizations, instead of verbs, is a common failing in academic writing. Examples of nominalized verbs are shown in table 14.2.

To illustrate, example 14.9 nominalizes every important action.

Example 14.9
Systemic infusion of fetal stem cells appears to be the most practical mode of administration; however, limited migration of cells to the target tissue may act as a constraint on its effectiveness.

The only verbs are "appears," "to be," and "act," which is sad, as there is no shortage of actions: "infuse," "administer," "migrate," "constrain," and even "target." We can convert many of those actions to verbs, tightening this sentence:

"The most practical way to administer fetal stem cells is to infuse them systemically; however, if cells don't migrate to the target tissue, this will fail."

Sometimes forcing the action into a nominalization pushes it out of a critical position in the sentence, as illustrated by example 14.10.

Table 14.2. VERBS AND THEIR NOMINALIZED EQUIVALENTS

Verb	Nominalization
Move	Movement
Differ	Difference
Suggest	Suggestion
Interact	Interaction
Analyze	Analysis
Develop	Development

In some cases, the verb and nominalization almost have the same form

Influence	A influenced B versus A had an influence on B
Approach	A approached the problem versus A took an approach to the problem
Yield	The reaction yielded a product versus The yield of the reaction was . . .

Example 14.10
Although models exist to calculate reaction rates as a function of molecular size, a failure to reproduce the experimental data is often observed.

This combines a nominalization with a passive to create a sentence with the minimum possible punch. The author is making an important point—these models often fail. However, that is nominalized to "a failure." This pushes the passive verb phrase "is often observed" to the sentence's stress, and it buries the critical action in the bowels of the sentence: "a failure to reproduce the experimental data." This would be better if the sentence's two clauses were effectively linked and if there were an active verb early in the second clause:

"Although models exist to calculate reaction rates as a function of molecular size, they often fail to reproduce the experimental data."

This works. It opens the second clause with the pronoun "they" to tie it back to the models, and then it hits the important point: "they often fail." This makes good subject–verb connection and puts the verb in the important place—the beginning of the main clause.

Another problem with verb nominalizations is that they are necessarily connected to fuzzy verbs. Because the action is named in the nominalization, and a sentence still needs a verb, it will be weak. Scan your work for nominalizations—there are probably more than you imagined. As a rule, turn them into verbs.

Table 14.3. ADJECTIVE NOMINALIZATIONS

Adjective	Nominalization
Different	Difference
Difficult	Difficulty
Able	Ability
Capable	Capability
Similar	Similarity

14.3.1. Adjective Nominalizations

There is another form of nominalization: converting an adjective into a noun. Examples of adjective nominalizations are illustrated in table 14.3.

Nominalizing adjectives also steals color and energy from writing. They leave it heavy and flat. For example, compare the following pair of sentences. Which is stronger?

Example 14.11
  A. The characteristics of this condition are the oxidation of membrane lipids, the denaturation of proteins, and a reduction in growth rates.
  B. This condition is characterized by oxidized membrane lipids, denatured proteins, and reduced growth rates.

Version A nominalized every adjective: "characteristics," "oxidation," "denaturation," and "reduction." In contrast, version B makes them all adjectives; the sentence is shorter and sharper.

Sometimes fixing a nominalized adjective can take several steps, as illustrated in example 14.12.

Example 14.12
  A. There was a difference between the reaction rates of treatments X and Y.
  B. Reaction rates were different between treatments X and Y.
  C. Reaction rates differed between treatments X and Y.

These all say the same thing with the action contained in some version of the word "differ." In version A, it's a nominalization—"difference"—and "was" is the only verb, a weak one. Version B is better, turning it into a real adjective—"different"—but it still uses the weak "were" as the verb. Version C puts the action into the verb "differed," and as a result it is both the shortest and most vigorous.

14.3.2. Why Do Nominalizations Exist?

If nominalizations are so horrible, why do they exist? Certainly, they weren't invented to clutter language, steal clarity, and make thoughts impenetrable! Naming something makes it concrete. Names hold magic. We use nominalizations to name concepts, which is useful. Could you imagine having to explain these ideas every time you used them?

Taxation without representation
Gene expression
Aromatic molecule
Ecosystem services
Epigenetics

Naming a concept is powerful because it defines a new schema, but it is also dangerous. It's dangerous because when you use the name, you assume that the reader knows and understands that schema. If your reader understands that an "aromatic molecule" is a ring with conjugated double bonds, you have effective shorthand for quickly and efficiently communicating a complex chemical concept. If they don't know the schema, however, and interpret "aromatic molecule" as "perfume," you can create some interesting miscommunication.

If your reader doesn't hold the schema, a nominalization becomes jargon—an unclear term that seems designed to exclude noninitiates from the club. With some audiences, you can safely use a nominalization, whereas with others you must define it. For the public, you would need to define "aromatic molecule" and would look arrogant if you didn't; for a paper in *Organic Chemistry*, on the other hand, you would look silly if you did.

The ability to nonimalize complex ideas also allows you to write sentences like "The arguments developed above . . ." In this case, "arguments" is a nominalization that encapsulates what may have been paragraphs' worth of text into a single word. That is powerful.

For a potent use of nominalizations, lets go back to example 12.1 from Winston Churchill: "until in God's good time, the New World, with all its power and might, steps forth to the rescue and the liberation of the old." Churchill put the nominalizations "the rescue" and "the liberation" in the sentence's stress. He could have made them verbs: "until in God's good time, the New World, with all its power and might, steps forth to rescue and liberate the old." This is weaker—the verbs don't have the same mass and solemnity, and Churchill deliberately left the action on "steps forth." He was encouraging the United States to step forth so that Britain wouldn't need rescue and liberation! Churchill cleverly used a tool to create eloquence. He also used parallelism and repetition (*the rescue* and *the liberation*) to add weight to his message, and drove it in by putting it in the sentence's stress. Churchill was a master of the English language; he knew when to break the rules, and how to use all the linguistic tools available to him. You might not save the world with your writing, but you might fund your graduate students.

Find the action in your sentences, put it in your verbs, and put them early in their sentences. If you do, your writing will be clear and lively. Sometimes a passive or nominalization will strengthen your writing, and sometimes they are essential. Every time you use them unnecessarily, though, you make your writing heavier and more opaque. A single unnecessary nominalization won't destroy your writing, but remember is wasn't the last straw that broke the camel's back—it was the accumulation of all the straws. Don't accumulate straws.

EXERCISES

14.1. Analyze published papers

Look at the papers you have been analyzing and read the critical paragraphs that define the opening, action, and resolution. Evaluate the actions and the verbs. Do the authors put the action in their verbs? Do they use active verbs? If not, try rewriting those paragraphs using stronger verbs.

14.2. Write a short article

Go back to your short article. Go through it sentence by sentence, noting the actions you describe and the verbs you use. Is every action in an active verb? If not, can you convert them into active verbs? If you choose to leave *any* action as anything but an active verb, justify your choice.

14.3. Revise

   A. Increased mobility of predatory nematodes in soil would increase opportunities for ecological interactions and so alter bacterial population dynamics.
   B. Polyaromatic hydrocarbons and polychlorinated biphenyls present enormous challenges in remediation, invoking large financial costs and presenting significant health risks to the workers who face exposure to the compounds.
   C. It was demonstrated that extraction of soils by NH_4Cl caused an enhancement in the recovery of Al relative to an extraction with K_2SO_4.

15

Words

> In literature, the ambition of the novice is to acquire the literary language; the struggle of the adept is to get rid of it.
>
> —George Bernard Shaw

As I said in chapter 11, words are to sentences what atoms are to molecules. They control the chemistry and "voice" of your writing—how it sounds and feels. Some atoms are inherently dense and toxic, like lead. With others, toxicity comes from their specific combination; carbon, hydrogen, and oxygen can produce fresh and fruity aldehydes but with just a slight tweak become rancid acids. So, too, with words. You can poison your writing with toxic words and toxic combinations.

Choosing words is not easy. English has amassed words from many sources, and different words convey different impressions of what you are saying and even of who you are. Just consider "fornicate" and its four-letter synonym, "f---."

Academics have an almost proverbial fondness for long, heavy words. Some use them because they think it makes writing sound more scholarly or because they want to show off their erudition, as Dennis Dutton, editor of *Philosophy and Literature* once accused the author of a notably convoluted piece of academic writing: "This sentence beats readers into submission and instructs them that they are in the presence of a great and deep mind. Actual communication has nothing to do with it."[1]

More of us, perhaps, learn to write in a heavy academic style because we imitate what we read and strive to "acquire the literary language." Over time this style becomes ingrained habit, creating a self-sustaining cycle. Writing this way also identifies us as members of the club, but one increasingly isolated from broader society.

The other reason we write "heavy" is because written and spoken English are different. We think differently when we write compared with when we speak. Written language is more formal, and our papers will outlast us, reinforcing a formal writing style. We lean toward longer and more elaborate words than we might otherwise choose.

But scientific writing can have life and energy—you can be professional without being pedantic. In earlier chapters, I've included examples of lively writing and discussed some ways to achieve that. The last method is to choose good words.

Written English is different from spoken English, but the difference should be primarily in sentence structure, not vocabulary. When you write a big word, ask yourself: "would I use it if I were talking to a friend?" For example, medical papers use language such as: "the therapy was efficacious." Education researchers write about "students with different learning modalities." Would you say either in normal conversation? I wouldn't. I would say, "the treatment worked" or "students with different learning styles." To most of us, these alternatives mean the same thing. Or maybe not—to some, "learning modality" might mean nothing at all, whereas "learning style" is clear as a bell.

Why not impress your readers with the sophistication of your vocabulary, showing that you can write technical-sounding language with the best of them? As a simple answer let me pose a question: when you last read a paper that was hard to read, were you impressed by how scholarly the authors were? Or were you frustrated trying to figure out what they were saying? We notice language when it's awkward, and may blame ourselves for not being smart enough to figure it out. When the writing is good, we notice the ideas and the data, and those are what make the science.

15.1. JARGON

Many describe science as filled with jargon, by which they usually mean arcane and uninterpretable terms that obfuscate our ideas. Naturally, most books recommend avoiding jargon as critical for clear writing. Yet science is technical and

1. Dennis Dutton, On Philosophy and Literature's annual "Bad Writing Contest," *Wall Street Journal*, February 5, 1999.

requires many specialized terms. When is a term avoidable jargon, and when is it a necessary and irreplaceable technical term? I distinguish them as follows:

Jargon: (A) A term that refers to a schema the reader does not hold. (B) A term for which there is an adequate plain language equivalent.
Technical term: (A) A term that refers to a schema the reader *does* hold. (B) A term for which either there is no plain language equivalent or where using it would be confusing.

This distinction is fuzzy and fluid and depends on the reader's knowledge. In one context, a word may be a technical term, but in another it may be jargon. If you use a term without defining it, it may be jargon, but if you define it in language understandable to your readers, you may transform it into a useful term. If you define a word that is well known to your readers, however, you may appear ignorant. A chemist would never define "mole" in a research paper, and a molecular biologist would never define "gene." Even high school science students should know those terms.

As an illustration of the fluid boundary between technical terms and jargon, consider the phrase "net primary production" (NPP). This is a measure of plant growth—the live biomass produced in an ecosystem. If I were giving a public talk and discussed NPP, the audience would be confused, so I would just say "plant growth." But if I said "plant growth" to an audience of ecologists, they would be confused—did I mean NPP, gross primary production, net ecosystem production, or some other measure of plant growth?

You need to use the terms that work for your audience. When you are trying to expand that audience, be sensitive to language and whether your technical terms are their jargon. Can you use simpler terms that will expand your audience without annoying the experts?

15.1.1. Avoiding Jargon

How and where you introduce a term may determine whether readers react to it as jargon. Remember the old/new information structure readers expect in your sentences? If you introduce a term in the topic position, readers interpret it as something they are supposed to know and are more likely to see it as jargon. If, however, you introduce a term in a sentence's stress, you dejargonize it. It will feel like you are defining the term, which might be good or it might be overkill. I illustrated this in example 12.3 about N mineralization.

What do you do with a term that might be too well known to put it in the stress but not well enough for the topic? See how the authors handle this in example 15.1; this is from a paper about the role of solvents in regulating the thermodynamics of chemical reactions. The authors discuss linear response theory, a theory most readers would probably know. However, the authors didn't make that assumption.

Example 15.1
This idea that excited states relax with rates determined by the solute-solvent system's ordinary energy fluctuations, commonly called linear response theory, is a critical component in the success of transition-state theories of chemical reaction rates in liquids.[2]

If you have studied basic chemistry, you should know that chemical reactions go through a high-energy transition state that breaks down into the final products; it should be an easy step to accept that the solvent can affect this transition. Voilà! You've learned what linear response theory is about. If you are a physical chemist and already knew the theory, this definition would merely feel like a comfortable reminder. This paper manages to reach out to and educate the broadest possible audience without alienating the core. I don't understand the details of the paper and the title is gobbledygook to me, but I do understand generally what it is about and why it is interesting.

Note how the authors achieved this balance—they use topic and stress positions to control where they introduce information. The topic introduces concepts that are known to any chemist. Then they put the theory's name in the stress of a clause set off by commas. By putting the name at the end of its own clause, they put it in a local stress position and give it some emphasis, but by putting that clause in the middle of the sentence they limit the emphasis, making it feel like a reminder, rather than a new definition. They effectively used Clark's 2-3-1 rule of emphasis that I introduced in chapter 12.

Introducing the term this way required a longer sentence (37 words) than most reading ease calculators recommend, but it actually made comprehension easier. Long sentences aren't necessarily bad—you just have to write them well, as Moskun and coauthors did. Example 15.1 was both clear and sensitive to the readers—excellent writing.

Here is another example of using the 2–3–1 approach of embedding potential jargon in a parenthetical clause (the 3-position) to remind you of the term.

Example 15.2
Programmed cell death, or apoptosis, is prominent in neural progenitors and appears to play an important role in the development of the cerebral cortex.[3]

These authors placed the word *apoptosis* in a short clause where it reminds readers of the term but doesn't feel like they are defining it for everyone.

2. Moskun et al., "Rotational Coherence and a Sudden Breakdown in Linear Response Seen in Room-Temperature Liquids," *Science* 311 (2006): 1907–11.

3. J. N. Pulvers and W. B. Huttner, "Brca1 Is Required for Embryonic Development of the Mouse Cerebral Cortex to Normal Size by Preventing Apoptosis of Early Neural Progenitors," *Development* 136 (2009), 1859–68.

Together these examples suggest a general pattern for using technical terms in different places in a sentence:

Beginning of the sentence: You assume that *every* reader knows and understands the term. You run the risk of it appearing to be jargon if they don't.

End of the sentence: You define a new term for everyone. You run the risk of appearing ignorant if it is already an accepted schema in the field.

Middle of the sentence: You assume that most readers know the term. You are also indicating that the term itself isn't critical to your story. You run the risk of people missing the term.

There is no single perfect place to introduce terms. You have to evaluate your audience and what they know. If you err, err on the side of overdefining. Any irritation an expert might feel at seeing a term defined unnecessarily would be slight and short-lived. The confusion a novice might feel at not having a term defined could be large and permanent—they might stop reading your paper.

15.1.2. Jargon and Acronyms

The worst form of jargon has to be undefined abbreviations and acronyms (at least you can look up words you don't know). In searching for examples, I occasionally ran into papers that had opening sentences like the following: "DCs are APCs that initiate immunity." In this sentence, DC stands for dendritic cell, a term used in the title, so I was able to figure it out, but APC was not defined anywhere in the paper. It was only by going online that I was able to figure out that it stood for "antigen presenting cell"; another definition for APC—"armored personnel carrier"—seemed unlikely. Tossing around a field's jargon is a fine way to show that you are part of the in-crowd, but you should be making your work accessible to the largest community practical. That is why the *Chicago Manual of Style* dictates that "terms must be spelled out on their first occurrence." Using an undefined abbreviation assumes that everyone who might ever read the paper already knows what it means. How likely is that? Most people reading a paper in immunology presumably knew what DCs and APCs are, but making the abbreviations the opening words of an entire paper excludes new readers, rather than reaching out to them. It isn't harder to write: "Dendritic cells (DCs) are antigen presenting cells (APCs) that initiate immunity."

Spelling out your acronyms and abbreviations the first time you use them takes a few more words but makes the paper easier for everyone involved. It won't offend an expert because you're not defining a term they don't need defined, and it will help the novices. The only exception to this rule is abbreviations that are so common that every reader knows them. You don't need to spell out DNA; your aunt knows what DNA is but would be baffled by deoxyribonucleic acid.

When we create acronyms and shorthand names, we almost always do it for our own convenience. Then we get so used to using our terms that we start to assume that they are obvious. They usually aren't. Remember principle 1 is to make the reader's job easy. Name things for their convenience, not yours. For example, if you studied two forests, one deciduous and one coniferous, you might label them DEC and CON, not ASP and HBR after the places you sampled. We have to learn many terms to do science—don't add unnecessarily to the list.

15.2. UNNECESSARILY TECHNICAL

Using jargon that readers don't know actively excludes them. A lesser evil is using terms readers do know, but where a nontechnical word would do the job more powerfully. Frequently this type of jargon results from being overly specific and as a result, undercommunicative. Consider the following example.

Example 15.3
Current models suggest that climate warming could release 200 times more nitrogen from soils than is taken up annually by terrestrial autotrophs.

This statement argues that the potential N release from soils is huge. But using the phrase "terrestrial autotrophs" weakens that message. Plants are a subset of autotrophs; others such as lichens and algae take up N as well. So 200× terrestrial autotrophs is actually a bigger number than 200× plants, but it would have been better to write "200 times the amount of nitrogen taken up annually by plants." The common word is more powerful—it engages a stronger schema.

The following is another case of adding words ostensibly to create precision.

Example 15.4
California has a Mediterranean climate regime, in which the heaviest storms occur when moist subtropical air is entrained by major Pacific storms.

The word I have an issue with here is *regime*. A regime is a pattern of conditions, but climate is a pattern of weather conditions (i.e., a regime). So a climate regime is no more than a climate! "Climate" for some, sounds too common, it's something everyone understands, whereas "climate regime" sounds technical. But that's the problem—it sounds like it means more than just climate and so it can be confusing.[4]

Example 15.5 shows a different reason for creating an overly qualified term. I think these authors were so caught up in the habit of avoiding action verbs that they created an elaborate nominalization to avoid it.

---

4. Climatologists use *regime* as a technical term, that is, a shift in the Pacific decadal oscillation is a regime shift, but that isn't how it's used here. This is putting on a more complex implication to clutter a simple term.

Example 15.5
This suggests that SRT may be a causative agent of chronic pain syndrome (CPS).

Why not say "SRT may cause chronic pain syndrome (CPS)"? We know that SRT is an "agent," so identifying it as one adds nothing. We get trained to think that noun expressions like this are somehow more specific or technical than action verbs, but they are not.

15.3. EMOTIONAL WEIGHT

Technical terms define the characters of the story—specific objects, organisms, and processes. Choosing them well is important. But it is also important to choose the words you use to describe what those characters are doing. Good choices can make them soar, bad choices can make them land, painfully.

It may seem surprising, but an important issue in choosing words in English is their origin. Academic English takes words from three main sources, Anglo-Saxon Old English, Norman-French, and Latin. As modern English was developing in the Middle Ages, Old English was the peasants' language, Norman-French the nobles' (brought in with William the Conqueror), and Latin the scholars'. That legacy endures. Anglo-Saxon words feel comfortable and casual. French words feel formal. Latin words inevitably feel like jargon; they were originally coined to show off the writer's education.

My field is soil science. *Soil* is from French; the Anglo-Saxon word, of course, is *dirt*. While people occasionally say "huh?" when I say I'm a soil scientist, at least they understand I'm an academic. If I say I study dirt, they are baffled—dirt seems too common to study. We call it soil science because we want to play on the positive connotations of the French word—soil grows plants, and the word has an elegant, flowing sound. Soil is good. That is why people use it as a euphemism: "the baby soiled its diaper" is a polite way of describing a messy event. Dirt, on the other hand, is what we get under our fingernails. The word is short, clipped, and one of the more generally negative words in the English language—calling something "dirty" is always an insult. Dirt derives from "drit," the Old Norse for "excrement," and it still carries a bit of the emotional legacy of that origin.

There are many times where we have a choice of French or Latin words and perfectly good Anglo-Saxon alternatives, as illustrated in table 15.1. Not only are the Anglo-Saxon words emotionally lighter, they also usually shorter. Even when both words came from French, the one assimilated earlier is generally shorter and feels more common.

Despite the benefits of short, light words, academics routinely fall into the centuries-old trap of choosing long, heavy Latin words. Many of us are still showing off instead of communicating. Given a choice of starting an experiment or initiating one, we go for the Latin and "initiate." Why use a long Latin word when a short Anglo-Saxon one will do the same job?

Table 15.1. EXAMPLES OF LONG FRENCH/LATIN VS. SHORT ANGLO-SAXON WORDS

Long French or Latin Word	Short, Anglo-Saxon Word (unless otherwise noted)
Duration (French.)	Length or time
Consume (French)	Eat
Mortality (French)	Death
Permit (French)	Let
Necessary (French)	Need
Demonstrate (Latin)	Show
Donate (Latin)	Give
Initiate (Latin)	Start
Attempt (French)	Try (from Old French *trier*)
Utilize (French)	Use (from Old French *user*)
Methodology (Latin combined form)	Method (Latin borrowed into English)

Example 15.6
We performed a study of six-months duration on the mortality rate of rats following exposure to elevated levels of lead. [20 words/119 characters]

Why write this sentence when you can write the following? "We did a six-month-long study of the death rate of rats exposed to high levels of lead. [18 words/84 characters]"

These sentences say the same thing, yet the second one is easier to read and 20 percent shorter as well. Proposals have page limits—you can't afford to waste space.

As a guideline, words ending in *-ate* are derived from Latin and sound heavy and full of themselves. Words ending with *-ion* are French. If you're not sure about a word, consult the *Oxford English Dictionary*—it gives the word's origin and meanings. It's worth noting that all the fuzzy verbs I listed in the last chapter are French or Latin. That is not surprising—common language is concrete, so when scholars reached for fuzzy verbs, they reached for Latin.

Sometimes, of course, you should use the French or Latin because the Anglo-Saxon word has a different connotation. Let me go back to example 14.6 about herbivores and exotic grasses. I suggested writing that sentence as: "Herbivores preferentially eat native plants." I think many would write this as "Herbivores preferentially consume native plants." because the Anglo-Saxon "eat" seems too visceral and too common to use in technical writing. Yet herbivores do, in fact, eat, and there is nothing wrong with saying so. In this context the words are synonyms, so use the shorter word. In other contexts, however, they are not synonyms and you could not switch them interchangeably; for example you can say that a fire "consumed the fuel," but not that it "ate the fuel." *Eat* implies mouths and nutrition, whereas *consume* carried its definition "to destroy" from Latin into English. Animals eat, fires don't.

Another alternative to saying "Herbivores preferentially eat" might be "Herbivores preferentially forage on . . ." *Forage*, however, carries the definition "to collect from" and so implies a hunting strategy, rather than a taste test. If that nuance is desired, use the longer Latin word, but be careful about relying on nuance; some readers may not understand the distinctions.

If you are struggling with word choice, a thesaurus is valuable, but you need to back it up with a good dictionary. So-called synonyms can have different implications, such as "consistent" and "coherent." My thesaurus lists these as primary synonyms for each other. Yet "consistent" suggests constancy, maybe even when it isn't desirable. In the words of Ralph Waldo Emerson, "A foolish consistency is the hobgoblin of little minds." So occasional inconsistency is desirable, but is it ever good to be incoherent?

15.4. PREPOSITIONAL PHRASES VERSUS COMPOUND NOUNS

A prepositional phrase, such as "rate of reaction" is made up of an object (reaction) and a modifier (rate) tied together with a preposition (of, in, on, etc.). The alternative is to use an expression such as "reaction rate" in which one noun directly modifies another: this is a compound noun (table 15.2). Prepositional phrases are usually nasty—longer and clunkier than the compound noun. They also have a strange attraction for nominalizations and passive verbs.

Table 15.2 lists some representative prepositional phrases and the alternative compound forms.

I'm not sure why so many people default to the prepositional over the compound noun. I think for some it sounds more precise. Others learned that compound nouns can cause problems (see below) and should always be avoided, as opposed to only avoiding them when they do cause problems. Others use them because we are being careless (as I often do in my first drafts).

Usually the compound noun is better, and for many things, we can't even imagine breaking them up—consider English without such expressions as "stone wall," "science fiction," or "Air Force"; or science without such terms as "benzene ring" or "nitrogen fixation." These expressions are short, clear, and effective ways of combining two things to build a more complex idea. You should generally turn prepositional phrases around to condense them, as illustrated in the following examples.

Example 15.7

A. The rate of the reaction increased sixfold when pH was decreased to 4.5.

B. The reaction rate increased sixfold when pH was decreased to 4.5.

Table 15.2. PREPOSITIONAL PHRASES VS. COMPOUND NOUNS

Prepositional Phrase	Compound Noun
Source of water	Water source
Supply of nitrogen	Nitrogen supply
Distribution of resources	Resource distribution
Kinetics of enzymes	Enzyme kinetics
Burning of fossil fuels	Fossil fuel burning
Cancer of the lung	Lung cancer

Example 15.8

A. This paper presents a new procedure for synthesizing complexes of iron and benzoate.

B. This paper presents a new procedure for synthesizing iron-benzoate complexes.

Example 15.9

A. Assembly is a stepwise process, starting with binding of Red22 to the coding region followed by binding of Red25 and Blu17 to the control region.

B. Assembly is a stepwise process, starting with Red22 binding to the coding region followed by Red25 and Blu17 binding to the control region.

In each case, the second version is a little shorter and a little tighter. In the last case, flipping the prepositional phrase turned "binding" from a nominalization back into a verb—a double win.

15.4.1. When to Leave a Prepositional Phrase

I've argued, as a principle, that every tool in English has value, and that includes prepositional phrases. So, when *should* you use one? As an example, consider the following sentence.

Example 15.10

These results suggest that modification of resource allocation allowed *Vaccinium* . . .

You could remove the prepositional phrase "modification of resource allocation," which would convert the sentence to the following: "These results suggest that resource allocation modification allowed *Vaccinium* . . ." But "resource allocation modification" is a jumbled mouthful of words, all the worse because they are nominalizations modifying each other. This is heavy, clunky, and hard to figure out. Such overdone compounds are sometimes known as noun clusters, but my colleague Ruth Yanai calls them "noun trains," a lovely term. Noun trains are worse than prepositional phrases. You can break them up into manageable pieces by using the occasional preposition.

How do you decide between a clunky prepositional phrase and a clunky noun train? If there are only two nouns, a compound is almost certainly better. If there are four nouns, break it up. Three is trickier; for example, "resource allocation modification" is awkward, yet "science fiction writer" is not. Several things make one a nasty noun train whereas another is fine. First is the complexity of the words: big words strung together form an undigestible mass. Second is whether we intuitively lump two of the words into a single unit—we read "science fiction" as one unit, so we see "science fiction writer" as only two units (a writer of science fiction); that's okay. We read "growth allocation modification" as three separate units and awkward.

Such extended noun trains can create confusion as to which is the core noun and which is the modifier. For example, is "Arctic system science" the science of studying the Arctic system, or is it system science done in the Arctic? The former focuses on the integrated system; the latter includes studies on individual systems that comprise part of the larger Arctic system. This is not a purely semantic debate—it has at times controlled major research directions and funding decisions by the National Science Foundation.

A noun train can even create confusion as to whether a word is a noun or a verb. Consider the expression "microbial community composition influences" in the following sentence.

Example 15.11
Current theory suggests that microbial community composition influences are most likely to be observed for physiologically narrow processes.

"Influences" is a nominalization, but it could be a verb, were the sentence "microbial composition influences soil processes." As someone reads the word, they will unconsciously assume one or the other. There is a 50 percent chance that they'll guess wrong and get pulled up short when they read the next word and have to back up and reinterpret. Any time you break the flow, you create problems. Here the prepositional phrase adds words but makes the idea clearer: "Current theory suggests that the influences of microbial community composition are most likely to be observed for physiologically narrow processes."

One final way you can use prepositional phrases is to control which word lands in a sentence's stress position. Consider the following two sentences:

Example 15.12
   A. Ecosystems can be managed to limit the effects of global warming.
   B. Ecosystems can be managed to limit the global warming effects.

In this case, the first sentence puts the strong phrase "global warming" into the stress, and is probably preferable.

As a last example, I want to go back to example 7.3 about signaling in visual transduction. That included the following sentence: "Despite the tantalizing evidence for DAG and/or its downstream products in visual transduction and the synergistic role of calcium, in no instance has application of such chemical stimuli fully reproduced the remarkable size and speed of the photocurrent."

I argued that these authors used topic and stress effectively to put emphasis in the right places. But look at what they had to do to put the stress on "remarkable size and speed of the photocurrent." They used the phrase "in no instance has application of such chemical stimuli reproduced."

Wow. A passive-feeling, nominalized, prepositional phrase—the verb is "has," and the action is the nominalization "application." That's a lot of no-no's packed into a mere six words. But it worked. This is a long, complex sentence, but its meaning is clear and it doesn't sound bad.

They could have written this as: "Applying such chemical stimuli has never fully reproduced . . ." That would have made the key word "application" into the verb "applying," but it would have put the critical word "never" in the middle of the clause. Instead, they put "never" up at the front of the clause to highlight it—they were using the 2–3–1 rule within a clause. Breaking some of the rules allowed the authors to put the right information in the right place to make the story flow.

This chapter covers only a selection of issues involved in choosing words to write clearly and engagingly, but it illustrates the principles. You are working to become an adept, so struggle to get rid of the literary language. Use the necessary technical terms, but avoid unnecessary jargon—and be aware of the difference! Remember that there are ways to remind readers of terms they may be unfamiliar with. Choose short, active words and phrases over long, ponderous ones. If you can do these things, your readers will be happy, and you may have more of them.

EXERCISES

15.1. Analyze published papers

Go to the papers you've been reading. Pick a paragraph or two and analyze the words the authors use. Go through each issue raised in this chapter and see whether you can lighten up the writing by avoiding jargon, picking shorter words, and eliminating prepositional phrases.

15.2. Write a short article

Go through your short article, and lighten up the words you use wherever possible. Can you do a stronger job of avoiding jargon, picking shorter words, and eliminating prepositional phrases?

15.3. Revise

Lighten up the following sentences:

A. The ability of animals to arrive at solutions to problems has been undervalued because studies have not been done that are considered to have scientific reliability.

B. Rats that had been maintained under varying environmental conditions demonstrated improved cognitive ability relative to the control group, which had been maintained in conditions that were invariant.

16

Condensing

Brevity is the soul of wit.

—Shakespeare, *Hamlet*

The Project Description may not exceed 15 pages.
—National Science Foundation Grant Proposal Guide

Shakespeare had it right: good writing is tight and sleek, giving you what you need with just enough extra to create flow and highlight. Bloated writing is bad writing.

For a scientist, writing compact English goes beyond being an act of style. It is an act of success or maybe survival. Papers rarely have a page limit, but when your ideas are buried in words, cumbersome sentences, and extraneous information, readers get confused and frustrated, potentially leading to extra rounds of revision or outright rejection. Proposals invariably do have strict page limits. The National Science Foundation won't look at a proposal that goes one word over 15 pages, but my first drafts are never under 18. Somehow those 18 pages *must* squeeze down into 15. How do you do that?

There are two approaches to condensing. The first is to tighten up your ideas and language. That's a skill which takes time to develop. But if you don't develop it, the only alternative is formatting tricks: using a smaller font, packing hypotheses into paragraphs instead of bulleted lists, and shrinking figures to the point you need a magnifying glass to read them. Though this does force all the words onto 15 pages, they end up looking like figure 16.1—a dense mass that a reader has to struggle through to figure out whether the ideas are worthwhile. Don't

Figure 16.1. A blurred page from a densely packed proposal.

think we don't notice. We may forgive you, but it's a hurdle to enlisting us as your advocate.

Some people who write like this argue that they have too much to say and can't condense the writing. Rubbish. I've *never* read something that looked like figure 16.1 where I couldn't streamline the ideas or condense the language without cutting substantive content. Never. Inevitably dense or densely packed writing means the author lacked either the skill or the inclination to condense, not that it couldn't be done. And they pay a price for it—rejection. By skipping the time to streamline your writing, you may feel that you are saving time, but in fact, you are squandering it.

16.1. A STRATEGY FOR CONDENSING

To write concisely, in *Writing Tools*, Roy Peter Clark gives the advice of "Prune the big limbs, then shake out the dead leaves." He elaborates by saying that "brevity comes from selection, not compression, a lesson that requires lifting blocks from the work."

"Prune, then shake" is excellent advice. First figure out what you don't need to say; then, don't say it. That's the "prune the big limbs" part. It grows directly from SUCCES and figuring out your simple story. Once you've figured out the story, you should be able to identify what information to include. The rest goes. I discussed this in chapter 3 (SUCCES) and chapter 8 (e.g., figure 8.1).

The next step is shaking out the dead leaves. That means cutting unnecessary words from the pieces that stay. Your first step should be to compact your ideas by building good story arcs. Broken arcs are inefficient. When you discuss an idea in multiple places, you almost always repeat things. Cleaning up the arcs lets you eliminate the repeats and the words used in transitions. Figure 10.2 illustrated how when there are three complete arcs, there are only 2 internal transitions, but when those arcs were fragmented, there were 13. In chapter 13, I pointed out how linking ideas and developing flow often requires adding words. Unnecessary transitions create waste.

After eliminating unnecessary material and condensing story arcs, the last step is to work with the delete key. Each word should do work; it should add content, clarify meaning, or provide coherence. Yet in almost every document, some words are slackers—empty adjectives, redundant modifiers, and other types of filler.

Most of us include a lot of filler. Stephen King[1] describes the best advice he ever got as "Formula: second draft = first draft – 10 percent." I count on trimming 20 percent from my first drafts. In learning to squeeze proposals into 15 pages, I even developed a game to help: look at every paragraph that has a word or two hanging on a bottom line and figure out how to cut enough to pull them up into the body of the paragraph. Killing one word may save a whole line.

1. Stephen King On Writing. Scribner; 1st edtion (October 3, 2000), Kindle Edition.

Score! This helped me develop skill as a literary trash compactor and to identify several targets for the delete key:

Redundancies
Obvious
Modifiers: adjectives and adverbs
Metadiscourse
Verbosity

16.2. REDUNDANCIES

Sometimes we use several words where one does all the work that needs doing. The next three examples illustrate this.

Example 16.1
I will develop, test, and apply a new synthetic approach to produce photovoltaic plastics.

Testing is part of developing, whereas "synthetic" and "produce" both refer to making things. So this sentence could easily read:

"I will develop a new approach to produce photovoltaic plastics."

Example 16.2
Most, but not all of the test subjects responded.

"Most" means "the majority" and so, "not all." This sentence could be written:

"Most of the test subjects responded."

Example 16.3
The effectiveness of these antibodies in HIV infection provides a proof-of-principle for the feasibility of using engineered antibodies as a novel therapy.

A proof-of-principle implies feasibility, so this can be condensed to:

"The effectiveness of these antibodies in HIV infection suggests the feasibility of using engineered antibodies as a novel therapy."

Often we repeat ideas in multiple sentences, in which case collapsing redundancy means collapsing sentences together. Sometimes we can delete an entire sentence, but often there is an idea, a few words, to capture.

Example 16.4
The altered precipitation patterns associated with global warming will change
the water regimes of most ecosystems, particularly those with arid, semi-
arid, and Mediterranean climates. These dry environments currently comprise one third of the terrestrial land surface.

Neither of these sentences is bad, but this can be collapsed down to:

*"Climate change will alter the water regimes of most ecosystems, particularly
those in arid and semi-arid regions, which comprise roughly one third of the
land surface."*

The original is two sentences because the authors' first was complex, so they
appropriately put how much of the land is dry in a separate sentence. But most of
the first sentence's complexity can be collapsed to two words: climate change.
Readers understand that includes both warming *and* altered precipitation; mentioning altered water regimes reinforces that. Simplifying the first sentence allowed
me to integrate the important point of the second into it and eliminate the transition words "These dry environments." I deleted "Mediterranean" because it is a
type of semi-arid climate, and it was only fleshing out the list. If the Mediterranean
climate were specifically important, it would be highlighted.

16.3. OBVIOUS

Obvious is close kin to redundant, as both encompass words that offer no useful
information. The difference is that whereas redundancies duplicate information
within a passage, obvious ideas are well known or implied and so don't need to be
said anywhere.

Example 16.5
There is evidence that X17-production can be associated with enzyme induction (Chu et al. 2008).

If there weren't evidence for the statement, the author wouldn't have said it, and
there certainly wouldn't be a literature reference. So "There is evidence that" is
obvious and can be deleted.

"X-17 production can be associated with enzyme induction (Chu et al. 2008)."

The author probably included the caveat "There is evidence" to suggest that
this finding is not confirmed. But the word *can* adds that caveat all by itself. If this
were an unequivocal statement, they would have written "X-17 production is
associated . . ."

Example 16.6
Snow cover is a characteristic of high alpine ecosystems that is critical in regulating both plant community dynamics and hydrology.

It's obvious that snow cover is a "characteristic" and alpine ecosystems are defined by being "high," so we can delete those and adjust a few words to fit the new structure:

"Snow cover in alpine ecosystems is critical in regulating both plant community dynamics and hydrology."

Example 16.7
The greatest challenge in dealing with the crisis of a pandemic is that it is global in scope and so public health responses must operate across national borders.

Two things define a pandemic: the disease is highly infectious and very widespread. Ergo, it is a crisis, and almost certain to be international. So, we can leave those ideas implicit without losing information:

"The greatest challenge in dealing with a pandemic is that public health responses must operate across national borders."

16.4. MODIFIERS: ADVERBS AND ADJECTIVES

Write with nouns and verbs. . . . The adjective hasn't been built that can pull a weak or inaccurate noun out of a tight place.
—STRUNK AND WHITE, *The Elements of Style*

The Adverb is not your friend.
—STEPHEN KING, "On Writing"

Adjectives modify nouns, and adverbs modify everything else (including adjectives). But good words don't need modifying. Strong, clear nouns and verbs give writing power, a power you can't match by decorating weak words. Eliminating unnecessary adjectives and adverbs will make your writing stronger and tighter.

Example 16.8
The entire reaction sequence takes less than one hour to complete.

Do you need both "entire" and "complete"? You could easily and condense this to:

"The reaction sequence takes less than one hour to complete."

You could even go further: *"The reaction sequence takes less than one hour."*

Example 16.9 illustrates using an adverb unsuccessfully to make a point.

Example 16.9:
The treatment dramatically increased X.

The author added "dramatically" to highlight that the increase in X was large. It doesn't work. "Dramatically" is fuzzy and doesn't carry much meaning— was the increase a factor of 2, 20, or 200? Without the concrete information on how much the treatment increased X, the adverb is weak. You could add that information:

"The treatment dramatically increased X by a factor of 42."

But if you know the increase was by a factor of 42, then it is obvious that the increase was dramatic. So just write:

"The treatment increased X by a factor of 42."

These examples illustrate what I call "empty amplifiers." They try to intensify the word they are referring to but don't add meaning (see table 16.1 for more). Empty amplifiers take up space but do no harm. If you delete them, the important thought remains. Take "dramatically" away from "dramatically increased X," and X still increased. Take "quite" away from "quite large" and you're still left with something big.

Modifiers can be more insidious than these empty amplifiers: they can hide empty thoughts. Sometimes when you strip away the modifiers, you find that there isn't a lot of substance left, as illustrated in example 16.10.

Example 16.10
The immune system uses a highly effective control mechanism that efficiently discriminates between self and nonself.

Table 16.1. EMPTY AMPLIFIERS : ADJECTIVES AND ADVERBS THAT TRY TO INTENSIFY THEIR REFERENT BUT ADD NO MEANING

With an "ly" these are adverbs; without "ly" they are adjectives.

Certain(ly)	Quite	Substantial(ly)
Dramatic(ally)	Rather	Very
Entire(ly)	Real(ly)	
High(ly)	Simple(ly)	

This sentence uses two adverbs and an adjective to emphasize how good a job the immune system does: it is both "highly effective" and discriminates "efficiently." If you delete those words, you are left with:

"The immune system uses a control mechanism that discriminates between self and nonself."

Without the modifiers, this sentence feels like it's missing something. We know what the immune system does, but this sentence could be rewritten to become more focused and concrete. What is the control mechanism? How is it efficient? That information may follow, but it should have been here.

This sentence replaced substance with hype and hoped we wouldn't notice. But decoration can never replace content. This sentence reminds me of the time I went to buy my father a bottle of Scotch for his birthday, and the guy at the store tried to sell me a simple blended in a cut-glass bottle, instead of the 25-year-old Macallan in a plain one.

This example was the opening for a story, setting up the picture that the authors fill in by illustrating how wonderful the control system is. So maybe it wasn't terrible, but it was an empty pawn push; the paper would be stronger with a queen launch that offered some intellectual meat.

Another example of using adjectives to create the sense that the author is saying something substantive is example 16.11.

Example 16.11
Thermal stress induces structural and functional changes in GTH-7.

Adding "structural and functional" appears to say something about the nature of the changes. But these encompass all possible types of change, changes that almost inevitably go together. So this really adds no concrete information. Deleting this phrase collapses the sentence to: "Thermal stress induces changes in GTH-7."

As with the previous example, this now seems like it doesn't say enough—it begs the question "what kind of changes?" But that question highlights the emptiness of "structural and functional." What we really want to know is the specific nature of those changes, and you're sure to tell us in the following sentences. Instead, integrate them into this one to make one tight sentence: "Thermal stress alters the conformation of GTH-7's active site, reducing its affinity for GXP."

16.4.1. Good Modifiers

The adjectives and adverbs I've discussed mostly reinforce the word they refer to and don't do enough useful work to justify their existence. However, some words don't just reinforce but clarify or define their referent. To illustrate, let's return to a modification of example 16.8. In the following, you could not delete the adjective "first."

Example 16.12
The first phase of the reaction sequence takes less than one hour to complete.

"First" distinguishes one phase of the reaction sequence from others; it provides essential information.

Some modifiers, rather than amplifying, alter the meaning of their referent. These are powerful. To illustrate this, Roy Peter Clark uses the example of "she smiled happily" versus "she smiled sadly." We expect smiles to be happy, so "happily" is an empty amplifier. We don't expect smiles to be sad, so "sadly" transforms the image entirely.[2]

A direct scientific parallel would be the difference between a "final result" and a "preliminary result." We assume results are final, so calling something a "final result" is wasting words. Describing something as a "preliminary result" suggests it's still tentative, a distinction that may be important. You could delete "final" but not "preliminary."

16.5. METADISCOURSE: TALKING ABOUT WHAT YOU'RE DOING

We often include some description of our actions and thoughts, rather than limiting our words strictly to the material at hand. For example:

We found that . . .
We argue that . . .
Our initial hypothesis was that . . .
These data may indicate . . .
To conclude . . .

This is known as metadiscourse—discussing the discussion. Some metadiscourse is necessary to develop the flow of an argument, but it can be obvious or redundant. Consider the following examples.

Example 16.13
We found that aniline did not react with . . .

These are new data, first reported in your paper—could someone else have found it? So write: "Aniline did not react with . . ."

2. You can argue that "entirely" here is an empty amplifier, but I like how it sounds and works. I want "transforms" as an active verb that directly follows its subject, but I also want to pull the idea of "transforms" into the stress. Putting the adverb in the stress achieves this. Remember: rules are guidelines.

Example 16.14
"In this study, we measured Y . . ."

This is a common expression but if you say "We measured Y," it's obvious that it was in this study. Going back to the idea of good adjectives, though, if you measured Y in a *previous* study, you would need to specify that: "In a previous study, we measured Y."

Avoiding unnecessary metadiscourse also eliminates concern about whether you should discuss your own actions in the active or passive voice. Some still object to saying "we found that aniline did not react" and insist on using the passive, leading to the cumbersome "aniline was not found to react." Eliminating the metadiscourse sidesteps the issue and produces text that is shorter and cleaner as well: "Aniline did not react."

16.6. VERBOSITY

I include verbosity as a separate category, but it is really the sum of multiple types of filler, creating sentences that ramble on endlessly. Verbose authors are often insecure, afraid to make a definitive statement, or can't separate their own mental processes from the story they are trying to tell. Example 16.15 is a particularly egregious case.

Example 16.15
The data show that some enhancement in the applicability of these measurements can be accomplished with freeze-fracture prior to analysis by laser-ablation mass spectrometry.

It's hard to characterize the junk that has been piled on this sentence, but if you cut it all out, the original 25-word sentence condenses to 11: "Freeze-fracture pretreatment improved analyses by laser-ablation mass spectrometry."

That is an example where the writing was awful and loaded with obese words. But verbose writing doesn't have to be terrible. Here is an example where the authors were trying to limit how much they packed into any single sentence.

Example 16.16
Maximizing the yield of X requires both optimizing the pH and selecting an appropriate catalyst. The optimum pH range is narrow, between 4.5 and 5, while appropriate catalysts include Mn and Fe.

This is structured as an LD story, with the first sentence describing the general conditions and the second detailing them. That would be fine if that detailing took an entire paragraph. But it doesn't. This can be cut in half: "Maximizing the yield of X requires a pH in the range of 4.5 to 5 and either a Mn or Fe catalyst."

16.7. VERBS AND ACTION

I mentioned in chapter 14 that active verbs are tight, while passives, fuzzies, and nominalizations are not—they require extra words. Putting the action into verbs is a powerful tool for condensing writing. As a reminder, I offer example 16.17.

Example 16.17
Agents that can interfere with the binding of AS2 protein to DNA are capable of delaying the onset of ovarian cancer.

By converting all the actions to verbs, this sentence becomes both shorter and stronger: "Agents that interfere with AS2 protein's binding to DNA can delay the onset of ovarian cancer."

16.8. A FULL PARAGRAPH EXAMPLE

The following paragraph is 260 words and I don't think it's terrible. I think it may be representative of decent first draft writing.

Example 16.18
A central dogma of ecology has long been that soil microorganisms must decompose organic matter, releasing inorganic N, before that N becomes available for plants to take up. In the arctic tundra, however, several lines of evidence have forced us to question the importance of microbial decomposition and inorganic N uptake by plants: 1) In these soils, microbes appear to take up enough inorganic N during the growing season that they leave inadequate supplies of N to support the N uptake needed to sustain measured plant growth. 2) The total annual net release of inorganic N by microbes is often half the value that is required to meet the demands of plant uptake, as estimated from plant harvests over the course of the growing season. 3) Several tundra plant species have been shown in lab studies to be able to take up amino acids from hydroponic solution, and can use the N to support growth. While these studies suggest that plants should take up amino acids and possibly other forms of organic N in the field, they do not provide conclusive evidence of this. Rather, amino acids are an excellent source of both C and N for soil microorganisms, which might be expected to outcompete plants for any free amino acids in natural soils, thus limiting the access of plants to these compounds. If tundra plants take up a significant amount of their N directly as amino acids, we must reevaluate our basic view of the central role of microbial breakdown of organic N to NH_4^+ in the tundra N-cycle.[3]

How much can we condense here without losing meaning? Let's work through it sentence by sentence.

3. This is adapted from early drafts of several proposals I have written.

~~A central dogma of ecology~~ *Ecological dogma* has ~~long~~ been that soil microorganisms must decompose organic matter, releasing inorganic N, before that N becomes available for plants ~~to take up~~.

Dogma is always "central," and implies long duration, so we can tighten the opening. Plants don't do anything else but take N up, do they? Obvious.

In the arctic tundra, however, several lines of evidence *challenge this.* ~~have forced us to question the importance of microbial decomposition and inorganic-N uptake by plants~~.

I deleted the metadiscourse, and instead of reiterating the decomposition/uptake concept, I encapsulated it in "this."

The next two sentences relate to microbial processes but would be unclear to a reader who is not a tundra ecologist. In the tundra, microbes take up inorganic N during the summer but release it during the winter. The distinction between "growing season uptake" and "total annual release," however, might be unclear to other readers. It needs to be either clearer or unsaid. The important point is that soil microbes don't release enough inorganic N to support plant growth—that is what challenges the dogma. So capture the best parts of each weak sentence to make one strong one.

~~1) In these soils, Microbes appear to take up enough inorganic N during the growing season that they leave inadequate supplies of N to support the N uptake needed to sustain measured plant growth. 2) The total annual net release of inorganic N by microbes is often half the value that is required to meet the demands of plant uptake, as estimated from plant harvests over the course of the growing season.~~
1) Microbes release only half the inorganic N required to support measured plant growth.

I took the idea of "plant growth" from the first sentence, because it is a stronger concept than "demands of plant uptake," but I took the core message from the second—it was stronger.

~~3)~~ 2) Several tundra plant species ~~have been shown in lab studies to be able to~~ *can* take up *and grow on* amino acids ~~from hydroponic solution, and can use the N to support growth~~.

This was unnecessary detail that would be in a reference. I eliminated the metadiscourse "have been shown" and collapsed the detailed explanation into the simple "can."

While these studies suggest that plants ~~should take up amino-acids and possibly other forms of~~ *use* organic N in the field, they ~~do not provide~~ *are not* conclusive ~~evidence of this~~.

By condensing "amino acids and other forms of organic N," to "organic N" I kept the important distinction of inorganic versus organic N. Condensing sharpens that message. In the last clause I cut the nominalization "conclusive evidence" to leave "conclusive" in the stress position.

> Rather, amino acids are an excellent source of both C and N for soil microorganisms, which ~~might be expected to~~ *should* outcompete plants for *them.–* ~~any free amino acids in natural soils, thus limiting the access of plants to these compounds.~~

The last phrase was implied—if microbes outcompete plants for amino acids, they necessarily limit plant access to them. This phrase may sound like it's adding information, but it isn't.

> If tundra plants take up a significant amount of ~~their N directly as~~ amino acids, we must reevaluate ~~our basic view of~~ the ~~central~~ role of microbial breakdown of organic N to NH_4^+ in the tundra ~~N-cycle~~.

Since we are reevaluating it, it is necessarily "our basic view." Nature doesn't change as a result of research—only our perception of it does. This is unnecessary metadiscourse. "Central" is an empty adjective. Now let's look at this condensed paragraph.

> Ecological dogma has been that soil microorganisms must decompose organic matter, releasing inorganic N, before that N becomes available for plants. In the arctic tundra, however, several lines of evidence challenge this: 1) Microbes release only half the inorganic N required to support measured plant growth. 2) Several tundra plant species can take up and grow on amino acids. While these studies suggest that plants use organic N in the field, they are not conclusive. Rather, amino acids are an excellent source of C and N for soil microorganisms, which should outcompete plants for them. If tundra plants take up a significant amount of amino acids, we must reevaluate the role of microbial breakdown of organic N to NH_4^+ in the tundra.

This is only 122 words: over 50 percent reduction with no loss of information. Because it carries less baggage, the message shines brighter.

16.9. CONDENSING TO CLARIFY

Sometimes you're faced with text that feels long and wordy but is also confusing. Often, the best way to clarify the message is to start by stripping away the excess to bring into focus what the text says (or doesn't say). Once you can see the message, it's easier to sharpen. As an example, consider the following passage.

Example 16.19
In the modern era of genomics, access to whole-genome sequence data is critical, but inadequate for the purpose of analyzing networks of physiological processes. The challenge is to effectively assimilate whole-genome sequence data based on objectively defined criteria in ways that facilitate interpretations and biological assessments.

The reaction most people have to this is simply: "Huh? What does that mean?" Is that reaction because the ideas are complex? Or because the language is? I argue the latter. This passage is rife with problems, including a massive deficiency of verbs (only four; can you find them?) and a surfeit of nominalizations. It violates almost every rule on word choice—so much so that the message is buried. To clarify, we can start by stripping this down to its bones. I underscore everything I think is filler and discuss it in the table.

In the modern era of genomics, access to whole-genome sequence data is critical, but inadequate for the purpose of analyzing networks of physiological processes. The challenge is to effectively assimilate whole-genome sequence data based on objectively defined criteria in ways that facilitate interpretations and biological assessments.

In the modern era of genomics	We know what era we're in—this is cliché and pompous. It's also a prepositional phrase.
purpose of analyzing	Verbose, nominalized way of saying "to analyze"
networks of physiological processes	Redundant and a prepositional phrase. Physiology implies process. This can be condensed to "physiological networks."
effectively	Empty adjective: of course we want to assimilate data effectively.
whole-genome sequence data	Repeats this phrase.
based on objectively defined criteria	Implied—we assume that criteria are objectively defined unless specified otherwise.
Assessments	Since we don't know what kind of assessments, this says nothing

If we clear all this away, this collapses down to the following, which is roughly 40 percent shorter.

Access to whole-genome sequence data is critical, but inadequate for analyzing physiological networks. The challenge is to assimilate the data in ways that facilitate biological interpretation.

A little work with the delete key improved this substantially. Not only is the point starting to emerge, but the writing feels more alive. By cutting clutter, it increased the verb-to-word ratio to 4 out of 27, a comfortable number. Doing this also exposes the second problem with this piece: most actions are nominalized.

1. What are the actions?
   <u>Access</u> to whole-genome sequence data is critical, but inadequate for <u>analyzing</u> physiological networks. The challenge is to <u>assimilate</u> the data in ways that <u>facilitate</u> biological <u>interpretation</u>.

2. Where are the verbs?
   Access to whole-genome sequence data <u>is</u> critical, but inadequate for analyzing physiological networks. The challenge <u>is to assimilate</u> the data in ways that <u>facilitate</u> biological interpretation.

The only actions that are verbs are "to assimilate" and "facilitate," and they are heavy Latin words that don't show action. The other actions are nominalized. These problems make the writing longer and more confusing. It is also depersonalized: to whom are the data critical? Who is challenged? There is no "old lady screaming" anywhere in this passage. To make this more compelling, we can put the action into verbs and use active voice.

To analyze physiological networks, we need whole-genome sequence data, but such data alone are inadequate. The challenge is to assimilate them in ways that allow better biological interpretations.

This now puts the actor, "we," in the topic position and follows immediately with the action verb: "need." Real people are going to analyze these networks, and those people are us; because the data are critical, *we* need them. I shortened "facilitate" to the lighter "allow." I could, perhaps, have shortened "assimilate" to "use," but when modelers talk about assimilating data into models it implies a suite of specific mathematical tools, whereas "use" does not. This is actually two words/three characters longer than the previous, but it reads more easily. It is much shorter and more intelligible than the original.

16.10. WHEN NOT TO KILL EVERY POSSIBLE WORD

In this chapter, I focused on how to cut every unnecessary word. You should apply those text-squishing tools to everything you write. There are, however, two caveats to these guidelines. First, especially if you are working with coauthors, it's alright to be a little verbose in your first drafts. It's easy to strip down overloaded writing. It's harder to go in the other direction and fill in the gaps in underloaded writing, which can be cryptic and disjointed. To figure out what is missing, you need to know the data and the story well. Advisors and coauthors, however, rarely know the story as well as you do, and they don't know what is in your head. Don't be afraid to give them enough that their main work will be trimming.

Second, words that build flow and coherence are not unnecessary. Any word that helps your readers understand your message does useful work. That is why I voted for keeping "entirely" in my discussion of empty amplifiers in section 16.5.1. To further illustrate the value of a few extra words to build flow, compare the following two pieces.

Example 16.20
A. Writing can get stripped down to the point that it becomes barren. Some filler can add flow, so use it, but carefully.
B. Writing can get too barren. Filler can add flow. Use it carefully.

I like the first. It's longer, but it reads well. The second is painful.

Learn to put your writing through the trash compactor. Once you have done that, work on finding the golden balance between over- and underloaded, between bloated and cryptic. When you hit that point, your writing will be tight and sleek, with grace, style, and power.

EXERCISES

16.1. Write a short article

Take your short article and condense it by at least 10 percent. In chapter 2, I suggested that the article should be between 800 and 850 words. Now it should be between 720 and 765 words. If you can go shorter, see how much shorter you can go.

Edit your partners' articles and edit the down at least 10 percent. See whether you would all suggest the same edits.

16.2. Edit: condense the following

A. Polyaromatic hydrocarbons and polychlorinated biphenyls are challenging to remediate: it costs a lot of money and it threatens the health of workers who are exposed to the compounds.
B. When expression of *Chla* and *Chlb* were compared, similar patterns of transcript abundance were observed in plants at different developmental stages.
C. Inherent resistance is an evolved response to living in environments that are constantly harsh. Inherent resistance doesn't require a plant to induce any specific physiological mechanism in response to an imposed stress. Rather plants that are inherently resistant are characterized by traits such as high root biomass, extensive chemical defenses, and relatively low maximum potential growth rates.

17

Putting it All Together: Real Editing

These ten or twenty lines might readily represent a whole day's hard work in the way of concentrated, intense thinking and revision, polish of style, weighing of words.

—Joseph Pulitzer

The greatest challenge in environmental toxicology is that contaminants rarely come one at a time—that oily sludge may also contain PCBs, mercury, and lead. Treatment that you would use for oil might not work on the mixture, or would leave you with a toxic residue that needs additional clean-up. Decontaminating the site requires a series of processes, each of which solves one problem but may create others.

Writing is the same. Rarely will you find a single problem by itself. Rather, they come in a convoluted mix. Broken story arcs travel with weak linkages; prepositional phrases with passive verbs; jargon with nominalizations. Disentangling such toxic writing (i.e., most of our first drafts) requires bringing all of our tools into use and requires multiple passes through a piece, each solving one or two problems and sometimes unmasking others in the process. Few of us are good enough editors to see all the way from clunky start to graceful end.

There are no simple rules for fixing multiple problems because each piece is unique, but there are some general approaches. I start with the big structural issues and then work down into finer and finer details of word use and style.

Get the story right, make sure the OCAR elements are in place, and then work on getting the language right. This leads to the sequence SCFL.

Structure: get the structure of the story into shape.
Clarity: ensure that your ideas are clear and concrete.
Flow: make the ideas flow, linking one thought to the next.
Language: make it sound good.

You won't, however, be able to revise a piece by doing just four passes, each dealing with one issue. They overlap, and solving one usually requires considering others. Even while focusing on structure, you can't ignore clarity and language. If the topic is unclear, it's hard to define the structure, and a lack of clarity often results from weak language. You may have to repeat the SCFL process several times. Clarifying your language may force you to rethink the structure of your arguments.

Here is a short example that is representative of a lot of first-draft writing and how a careful writer uses the SCFL approach to convert it into a polished final draft. This is the opening to a paper on how plants compete under varying light levels and how that affects community composition. You can see how this author revised the first draft into a strong finished piece by considering multiple aspects of their writing even while focusing on one of the SCFL elements.

Example 17.1
Plant behavior in response to dynamic resource availability, such as changing light regimes, has been well studied. Adjustments made at the organelle to canopy level to modify light interception exemplify how responsive plants are at a range of scales to their resource environment.

17.1. STRUCTURE

Because this a short opening to a paper, the most important structural issue is to ensure that it accurately frames the paper's direction. A secondary goal is to make the sentences fit together.

The first sentence defines a story arc about plant behavior, the sentence's subject. But the verb is "has been studied," a passive verb that doesn't appear until the end of the sentence and is badly separated from its subject. Worse, this verb makes the sentence's point unclear. Is it that plant behavior has been studied, so this is about the history of science? Or is the story about biology—what those studies revealed? To define the structure for the paper, the authors needed to clarify; then they could rewrite to capture either of those story lines.

A. History: *Many studies have examined plant responses to varying resource availability.*

B.   Biology: *Plants use a range of approaches to respond to varying resource availability.*

Each of these options uses an active verb that directly follows its subject; they capture the directions this story could go. The author moved forward with B—the paper is about plant biology.

Note that they dropped the last clause, "such as changing light regimes," to tighten this sentence and leave the important idea of "varying resource availability" in the stress position. The point of this first sentence is to make the general statement introducing the story. This is a pawn push, positioning move.

17.2. CLARITY AND FLOW

Now look at the second sentence. It doesn't mesh well with the first, raising structural issues, but it also lacks both energy and clarity because of the language. Clarity is the biggest problem; the actions are that plants "adjust" and "respond" to varying resource levels, but those aren't expressed in verbs. Rather, the first is in the nominalization "adjustments" and the second in the adjective "responsive." The verb is "exemplify," which misses the real story. Because of the weak language, the sentence isn't concrete—what are the adjustments that occur at the organelle and canopy levels? The reader wants to know not just *where* the adjustments are but *what* they are.

A lesser clarity issue is that the grammatical subject ("Adjustments . . . interception") is 12 words long, so long that the verb is lost in the depths of the sentence. Technically, this isn't poor subject–verb connection because the verb actually does immediately follow the subject. But it feels disconnected because the subject is so long, and that is what matters to the reader.

In terms of structure and flow, look at the topic-stress–topic-stress linkage:

Plant behavior . . .	has been studied ←→	Adjustments . . .	resource environment
Topic	*Stress*	*Topic*	*Stress*

There is no connection between the stress of the first sentence and the topic of the second, even though the second is clearly supposed to illustrate the general point made in the first. These have to connect. In attacking the second sentence, therefore, the key issues to address were:

Clarity: Activate the verb and fix the subject–verb connection. Make the action concrete: what are the adjustments?

Flow: To connect to the opening sentence, make the topic relate to "a range of approaches to varying resource availability."

To do these, the author moved the ideas of "exemplify" and "varying resource availability" up front and connected the subject (plants) to a clear active verb (modify).

For example, as light varies, plants modify light interception by mechanisms that range from organelle-level physiological shifts to canopy-level changes in architecture.

Starting with "For example" makes it clear that this illustrates the first sentence's general argument. That allowed dropping the phrase "such as changing light regimes" from the first sentence, sharpening and shortening the passage.

This sentence then moves on to "as light varies," capturing the theme of "varying resource availability" that was the stress of the first sentence. The connection between the two is now solid, linking stress to topic.

The author made the sentence concrete by connecting the subject and verb—"plants modify" and briefly identifying the types of modifications at each level—physiology at the organelle and architecture at the canopy (how tall? how does it branch?). That isn't a lot of information, but it is enough to feel concrete.

17.3. LANGUAGE

At this point, the structure, clarity, and flow problems were fixed, but the language was still clumsy. The sentence repeats expressions and the prepositional phrase "changes in architecture" felt awkward. The author couldn't simply eliminate that phrase because then it would become "canopy-level architecture changes," which is a bad noun train. So the author tried another approach, one that took all the words that expressed change: "shifts" and "changes" and condensed them into one.

For example, as light varies, plants modify light interception by mechanisms that range from shifts in organelle-level physiology to canopy-level architecture.

This left an awkward phrase, "mechanisms that range from," the prepositional "shifts in," and repeated the words "light" and "level," all of which made the sentence clunky. One more round of editing fixed those issues.

For example, as light varies, plants modify its interception by mechanisms working at different scales, from organelle physiology to canopy architecture.

This eliminated a redundant "light" by using the possessive pronoun "its" and eliminated the awkward "mechanisms that range from" and the repeated "scale" by rewording it to "mechanisms working at different scales." This sentence is now concrete, identifying conceptually what the mechanisms are. It also has some music to it, ending with a nice parallelism in the stress position. The authors achieved these by packaging the overall story into three compact little arcs:

1. "For example, as light varies,"
2. "plants modify its interception by mechanisms working at different scales"
3. "[those scales range] from organelle physiology to canopy architecture."

Note that each phrase carries out a distinct OCAR function: the first opens, the second describes action, and the third resolves by telling you what the mechanisms are. Even within a sentence, getting structure right solves many problems and makes the rest much easier.

Through editing and attending to the SCFL and OCAR issues, this gained clarity and grace while losing words. What it kept, however, is professionalism. Clarity and grace didn't come from dumbing down, but from sharpening up, that is, good writing.

At this point, the sentences were strong and connected well, so it was time to cycle back to the beginning of the SCFL checklist to revisit structure. This raised a fresh question: if the paper is only about responses to varying light, why not kill the first sentence? It's just pushing a pawn, so skip it and launch a queen.

As light varies, plants modify its interception by mechanisms working at different scales, from organelle physiology to canopy architecture.

Now this was ready to roll! This piece started with 42 words, and it ended up with only 19, not one of them wasted.

This multistep SCLF process should be the norm in your editing. You will always start with something rough that you have to polish, just as the authors of this piece did. They produced a terrific third draft; unfortunately, many people don't. Get used to how much work it can be to take even a short piece like this and polish it into powerful prose. You fix one set of problems and expose another. So edit it again. And again. Until it reads effectively and gracefully.

There is one final secret weapon in revising, but it is best done in privacy. Close the door to your office. Print a clean copy. Clear your mind. Now, stand up, step away from your desk, and *read it out loud*. Awkward expressions, breaks in flow, clunky words—your eyes may skip over them, but not your ears.

The need to take multiple passes through a piece, fixing problems as they emerge, explains a frustrating phenomenon I experienced with my advisor, my students have with me, and you probably have as well. You write a draft, someone edits it, and you make those changes. Then, they edit their edits. Sometimes back to the way you had originally written them! Why didn't they get it right the first time? Are they just changing things to change things? Probably not. Every time you come back to a piece, you need to look at it afresh. Sometimes the changes you scrawled on a sheet of paper or typed in seem okay but don't really work. You may only realize this when you see the whole new piece or when you read it aloud. Sometimes changes elsewhere in a paragraph mean that you need to rewrite a specific sentence to fit the new structure. Remember the section on "Writing versus Rewriting" in chapter 1. Writing is a process of experimentation and revision; there is no single "right answer." My last word of consolation on this is that the more you do it, the easier it becomes. It might even become fun.

EXERCISES

17.1. Write a short article

Take your short article (or some other project you're working on) and apply the SCFL approach to analyzing and revising it. Take one of your peer's short articles and apply the SCFL approach to editing it.

18

Dealing with Limitations

No plan survives contact with the enemy.
—Helmut von Moltke the Elder

No research is perfect. Every study has limitations, and every data set has blemishes. You have to address these even while focusing on meaningful results and that understanding that grows from them. The question is: how do you address the negatives without undermining the positives?

Many writers address limitations using what I call a "yes, but" strategy, where they present the results and conclusions and then discuss their limitations. The most classic form of this is the "more research is needed" resolution. The problem with "yes, but" is that it puts the "but" last, making it the resolution and most powerful message.

So what is the alternative to "yes, but"? Turn it around. Instead of "yes, but," use "but, yes." Deal with the limitations as early as possible, get that discussion out of the way, and then get on with developing a strong story. Constrain your conclusions to fit within the limitations but end with a "yes."

"But, yes" allows you to end powerfully, making for better storytelling. Importantly, though, it is also easier for the reader to follow and, I believe, more honest. It achieves these aims by airing your dirty laundry up front, rather than having it feel like the fine print in an advertisement, something you hope readers will miss. Having laid out the limitations, readers can work through your data and discussion in light of them. They never have to back up and reprocess information—that takes work and creates confusion. Depending on the issues, you may need to address limitations in any part of a paper or proposal; as a rule, earlier is better.

18.1. THE INTRODUCTION: PROMISE THE STORY YOU WILL DELIVER

Many problems arise not from inherent limitations in the methods and resulting data but from a mismatch between the question and the methods. Authors ask a good question but use methods that are inadequate to answer it. Such problems can sometimes be resolved simply by revising the question. In the Introduction, frame the knowledge gap you will actually fill—set up expectations you can deliver on.

As an example, imagine a paper submitted to *Soil Biology & Biochemistry* in which the Introduction said the following.

> Example 18.1
> To determine the size of the bacterial population in soil, we plated bacteria on nutrient agar and counted colonies.

I would probably reject this paper without review. We've known for a century that at most 1 percent of soil bacteria grow on agar plates, so this method can't possibly determine how many bacteria there are in soil. To count total bacteria, you have to do microscopic counts or DNA analyses. If on the other hand, consider if the paper said the following.

> Example 18.2
> To determine the size of the bacterial population in soil that is able to respond to inputs of fresh, high-quality organic materials, we plated bacteria on nutrient agar and counted colonies.

For this question, the approach is at least defensible. In fact, the methods that would work for the first question (microscopy and DNA) would fail for this one. Those methods wouldn't tell you which bacteria would grow if you fed them; plating might. So this one I would send out for review.

Of course, if this were the challenge, we would have to revisit the opening to ensure they frame a knowledge gap about bacteria that respond to inputs, rather than about the total population. That redefines the research and constrains its scope, but doesn't necessarily make it less interesting or important. It might even make it more interesting—I'm fine with small insightful questions. As I've said before, it is better to fill a small knowledge gap than none at all, which is what readers will think if questions and answers don't match.

18.2. MATERIALS AND METHODS

The second place you can address limitations is in the Materials and Methods. When limitations relate to experimental details or analytical methods, discuss them immediately to lay any concerns to rest. If you don't, readers may decide the work is so flawed that there is no reason to read further, or they may be distracted enough by their concerns to miss your important points. At best, they remain skeptical and you would struggle to gain their full acceptance. You are much better off if you can address readers' concerns as soon as they arise.

An excellent example of how to do this is in a paper evaluating overwinter survival of snowshoe hares in northern Canada. One of the experimental treatments involved a large fenced exclosure to keep out predators.

Example 18.3
Our final two treatments increased hare survival by excluding predators. One 1-km² area was enclosed by chicken-wire and electrical fencing (predator-exclosure) from 1987 to 1996 to exclude terrestrial predators, primarily lynx (*Lynx canadensis* Kerr) and coyotes (*Canis latrans* Say).[1]

The limitation here is that the design is unreplicated—they had only one exclosure. Traditional statistical logic might argue that without replication it is impossible to determine whether effects were due to the exclosure itself. Some people might read this and think "unreplicated and unstatistical, therefore unreliable and unscientific." How to address this? The authors did so with this paragraph.

Example 18.4
The predator-exclosure and predator-exclosure + food sites were not replicated, owing to the high costs of maintaining 8 km of electric fencing on a nearly daily basis for a decade. Ecologists are caught increasingly in the tension between the manipulations necessary to understand large-scale ecological processes and the impossibility of replicating large manipulations. Our control sites sampled a wide range of natural variation within our study area, increasing the likelihood that differences on the treatments were due to our manipulations rather than to underlying natural variation.

This addressed nonreplication in a way that was effective and unapologetic. First, the authors made it clear why it would be impossible to replicate. Then they grounded this within the larger discussion of how we study ecology: we often can't use nice, tidy replicated experimental designs to study messy nature. Environmental scientists must (and do) accept pseudo-replicated experiments as long as the investigators are sensitive to and deal appropriately with their limitations. Inferences of causality have to go outside simple statistics. After all, we study

1. K. E. Hodges, R. Boonstra, and C. J. Krebs, "Overwinter Mass Loss of Snowshoe Hares in the Yukon: Starvation, Stress, Adaptation, or Artifact?" *Journal of Animal Ecology* 75 (2006): 1–13.

"earth system science" but have only one Earth. Finally and importantly, they discussed how they dealt with the limitations to convince readers that differences between control and exclosure were likely real treatment effects. Having done this, they didn't raise the issue again. They got on with discussing the real science and the important results.

Addressing methodological constraints as soon as you describe the method is even more vital in proposals than papers. When your audience is unsympathetic and critical, you can't afford to give them any excuse to "go negative." Particularly when people disagree about methods, it is important to explain why yours will answer your questions. If there are limitations, explain how you get around them. If there are things that you are not going to do, tell us why (too expensive, unnecessary for your specific questions, whatever). If you avoid mentioning the negatives, reviewers will find them anyway, criticize you for them, and probably recommend rejection. If you discuss them openly, a reviewer may still object, but you might convince them you are right, or at least that you've thought about it enough to get the benefit of the doubt.

An example of this strategy is a proposal I wrote some years ago to characterize soil organic matter (SOM) in the arctic tundra. Tundra soils contain a huge stock of organic C that could be released as CO_2, accelerating climate warming. We wanted to find out the amount and chemical form of the SOM that is biodegradable. The challenge was that SOM is comprised of a complex mix of molecules, ranging from fresh plant sugars to ancient humified gunk. Some of these are easy to process, whereas others are effectively inert. There are no methods for unequivocally characterizing these materials. There are, however, several methods that give insight into specific aspects of SOM chemistry.

The first time I submitted the proposal, I described the methods without discussing their challenges. Reviewers hit on the methods' limitations, and the proposal was rejected. In the resubmission, I included the following section (slightly condensed).

Example 18.5
Potential pitfalls
Each of the methods proposed has potential limitations, and the utility of several have been questioned. However, we believe that we have constrained the limitations adequately. We have done this by being careful in interpreting what the data actually tell you (e.g. isotope equilibration as a method for estimating microbial metabolites rather than the total active fraction) and in part by carefully integrating different approaches to complement each other. Thus we believe that this suite of methods will do a good job of characterizing the nature of available C and N across the tundra landscape.

The isotope equilibration technique relies on several assumptions. As originally construed, it required the assumption that it labels the active fraction without labeling the recalcitrant fraction. This is debatable since some microbial products are recalcitrant. However, the assumption that it labels the microbial metabolite pool is substantially more robust. There is rarely extensive abiotic immobilization of either $^{15}NH_4^+$ or glucose, so the isotopes should be in the metabolite pool.

The utility of chemical fractionation in characterizing soil organic matter has been debated ever since the first fractionations of fulvic and humic acids. Chemical fractionation of *plant material*, however, has been quite successful in characterizing litter quality. Since in tundra soils much of the organic matter is still plant detritus these techniques should be useful.

The second time around, the proposal was funded. Reviewers still commented on the methods and the challenges of the project, but enthusiastically, recognizing that we had balanced methods' strengths and limitations and we had thought about how to interpret the results. We didn't change the methods, just how we defined their results.

I sometimes say that there is no such thing as a bad method. As long as it is done well, a method gives solid data and real information. The "bad" part always relates to interpretation—a method may truly be bad for measuring one thing but good for something slightly different. In Example 18.3, the isotope equilibration technique was bad for estimating the active fraction, but good for estimating microbial metabolites. Make sure to explain why your methods give the information you need.

18.3. DISCUSSION

When constraints are methodological detail, address them in the Methods. When they go beyond that to affect how you interpret the data, you must address them in the Discussion. You must openly discuss limitations but without highlighting them so strongly that they become an argument for rejecting the paper entirely. Doing this involves both story structure and language.

In terms of structure, the question is where to put the discussion of limitations. Most of the time, you should avoid the power positions of the Discussion's opening and resolution. Those are for critical story points, and if limitations are the critical point, why submit the paper? You should generally find some convenient place early in the body of the Discussion to discuss the work's limitations and constraints.

As for language, do it briefly and honestly, then get on with the story. Example 8.7 used this approach. The authors included two constraints on the conclusions, but they put them in the middle of an LD paragraph—after the lead but before the rest of the discussion.

A good discussion of methodological limitations comes from a paper that explored the side effects of inhaled cortical steroids (ICS) as a treatment for asthma. Clinical trials may underestimate side effects because trials are usually short and have restricted selection criteria for who participates. To better evaluate side effects of ICS, the researchers surveyed people who routinely used them and found effects ranging from gummy mouths to mood swings and sleeping problems, some of which had not been well documented. But survey-based studies always have limitations that the authors rightfully felt they needed to bring to readers' attention.

Example 18.6
Limitations of our research include potential selection bias. Although our response rate (56%) is not atypical for a questionnaire study, we acknowledge that responders may have had a greater propensity for perceiving ICS side effects as compared to non-responders, and it is possible that side effect reporting may have been inflated in all dose groups as a result. Alternatively, ICS users in our study may have experienced relatively mild ICS side effects, since they continue to use ICS regardless of their perception of side effects.[2]

This pointed out that there is some quantitative uncertainty about the results that can never be overcome in a survey study—you can't make people answer, or answer honestly. But there were robust patterns in the results, offering information that physicians should be aware of, which the authors then elaborated.

Another example of dealing with limitations in the Discussion comes from a paper about how variation between individuals can affect transcription patterns in different tissues, ultimately leading to different physiology.

Example 18.7
Despite the limited number of validated examples, osteoblast-specific transcript isoforms are present and indicates that expression of certain alternatively spliced variants in bone tissue are controlled by different regulatory SNPs and vary between individuals and/or populations, with potentially important biological functions.[3]

The authors start with the opening clause "Despite the limited number of validated examples," which frames the "but"; they then give the "yes" by saying that osteoblast-specific transcript isoforms *are* present and suggest the implications of that. They use the "but, yes" approach within a single sentence. They remind us of the limitations but don't dwell on them.

2. J. M. Foster, L. Aucott, R. H. W. van der Werf, M. J. van der Meijden, G. Schraa, D. S. Postma, and T. van der Molen, "Higher Patient Perceived Side Effects Related to Higher Daily Doses of Inhaled Corticosteroids in the Community: A Cross-Sectional Analysis," *Respiratory Medicine* 100 (2006): 1318–36.

3. Kwan et al., "Tissue Effect on Genetic Control of Transcript Isoform Variation," *PLoS Genetics* 5 (2009): e1000608. DOI:10.1371/journal.pgen.1000608.

18.3.1. An Initial Methodological Considerations Section

Although you should generally avoid raising constraints and limitations in power positions, on rare occasions it is necessary. When experimental approaches are unusual or novel, it can be important to give readers some calibration on their strengths and weaknesses before discussing the results themselves. In these cases, it is useful to open the Discussion with a section that weaves the methods considerations into the opening of its story arc.

An example of this approach is a paper I wrote evaluating controls over nitrogen movement between decomposing leaves in forest litter. As leaves fall, they form a layer on the soil surface, a layer that might include leaves from different tree species, some of which are N-rich while others are N-poor. This creates a complex environment in which the movement of N from rich to poor leaves may regulate the decomposition of the whole litter layer.

The experiment used an ^{15}N isotope tracer to follow N movement from ^{15}N-labeled leaves to unlabeled leaves, but it used the "microcosm" design shown in figure 18.1: two leaves (one labeled, one not) held together with wire and wrapped in plastic to prevent them from drying out—not quite what happens in an actual forest. Because it was a horribly artificial experimental design, we felt it important to be explicit about the constraints this placed on data interpretation, and we put that discussion in a prominent place that a reader couldn't miss. We started the discussion with an "Experimental Considerations" section.

Example 18.8

Experimental Considerations

This experiment used a rather artificial structure: pairs of individual leaves held together to ensure contact and minimize diffusion limitations. We labeled [with ^{15}N] only the microorganisms growing on the litter and so

Figure 18.1. The set-up for an experiment evaluating N movement between decomposing leaves.

were evaluating the short-term movement of N as regulated by the release and uptake of N by communities growing on different substrates. Although fungi were likely involved in decomposing the leaves, we did not observe fungal mycelia developing between them; such mycelia are thought to be important in moving N through a natural litter layer. Thus, this design provides little information about the amount of N that would actually move within natural litter layers decomposing *in situ*. However, this controlled experimental design allowed us to do a ^{15}N budget on each individual leaf-pair microcosm and thereby evaluate the bi-directional N movement among leaves of different species and N-levels, and to evaluate the separate role of source and sink N concentrations in regulating N movement within the microcosms.[4]

This was honest about the limitations of the experiment and that it told us little about the rates of N-movement in the field. But it provided a venue for arguing that we could assess mechanisms that controlled movement and that this offered a useful insight. This discussion also allowed us to remind readers of the knowledge gap we were targeting and the questions we were asking.

18.3.2. Within the Conclusions

You should never make limitations *the* conclusion, but sometimes you may need to mention them *within* the conclusions. In such cases, you need a tightly defined "but, yes" structure to frame the limitations as quickly as possible and then to push on to the conclusions—what you learned.

A particularly clever approach to doing this is illustrated in example 18.9. This is from a paper that evaluated approaches to estimating the mortality of children under five years old in nations that don't keep records of births and deaths. There was no absolute data set to use as a reference to assess the reliability of the estimates. That left uncertainties that constrained the conclusions. But if the authors had included that entire discussion in the concluding paragraph, it would have overwhelmed it. Instead, they discussed the limitations in the next-to-last paragraph, which was long and involved (21 lines). The final conclusions paragraph was only 11 lines.

Example 18.9
Despite these limitations, the methods proposed here represent a major advance on current practice and offer the prospect of vastly increasing our knowledge about levels, recent trends, and inequalities in child mortality. If we are to make rapid progress with the unfinished agenda of reducing child deaths, policy and practice must be better informed by more comprehensive,

4. J. P. Schimel and S. Hättenschwiler, "Nitrogen Transfer between Decomposing Leaves of Different N Status," *Soil Biology & Biochemistry* 39 (2007):1428–36.

relevant, and timely information. Systematic application of the methods proposed here will establish that evidence base, and thereby increase accountability among countries and the global health community to accelerate efforts to reduce the global toll of child deaths.[5]

This reminds us of the constraints, but with a tidy nominalization ("these limitations") that refers back to the previous paragraph. This simultaneously highlights and minimizes them! By separating the negatives and the positives into separate story arcs and using effective language to link them together, the authors addressed the constraints honestly in a way that did nothing to undermine the power of their conclusions. This was nicely done.

As I said at the beginning of this chapter, no research is perfect, and there is nothing to be embarrassed about in admitting it. Rather, the opposite is true—it's our responsibility to address our work's limitations. The "but, yes" approach does this in a way that is open about the limits but highlights the conclusions. It makes the presentation clear, strong, and credible. By clarifying both the limits and the power of your work, it will motivate others to pick up and build from it, that is, to cite it.

5. J. K. Rajaratnam, L. N. Tran, A. D. Lopez, and C. J. L. Murray, "Measuring Under-Five Mortality: Validation of New Low-Cost Methods," PLoS Medicine 7 (2010): e1000253. DOI:10.1371/journal.pmed.1000253.

19

Writing Global Science

Life isn't fair.

Science is increasingly dominated by scholars for whom English is a second language and by nations with developing scientific cultures. These don't necessarily overlap: Germany was a founder of the modern scientific tradition, whereas India, where English is well established, has a developing science program. Language and scientific culture, however, can each pose challenges to publishing in the international marketplace of high-impact journals.

Many people understandably feel language is *the* struggle. Writing a scientific paper seems daunting when even ordering dinner at a restaurant can be a trial. If English is your second language, you may feel that writing science must be easy for native speakers. If that were true, I wouldn't have written this book. Writing is hard for all of us.

The hardest part of writing science, though, is developing the story and laying it out cleanly. The essence of getting the story across is structure: knowing what to put where. Structure comes before language in the SCFL formula I discussed in chapter 17, and most of this book is about structure. Story structure transcends language; OCAR isn't about English.

Ultimately, the greater challenge is learning to do the kind of science that leading journals are looking for. That was a challenge for me, and I had teachers who were the academic offspring of generations of leaders. Figuring out how to be competitive in a sophisticated game without world-class mentorship is tough. Look at how long it has taken the United States to learn to play soccer.

19.1. DOING GLOBAL SCIENCE

To get papers published in an international English-language journal, you must structure an effective story and write it in correct English. But the first and most important step, of course, is to do good science. Learning to do good science is hard—you have to stretch your intellect and creativity to push the boundaries of knowledge. To develop those skills, most of us need training and mentorship, yet many places are still in the process of developing a culture of inquiry that supports and trains researchers to take risks, challenge established ideas, and question authority.

Doing science is inherently an act of both confidence and humility. Confidence in developing your own ideas and data, doing the work knowing it may fail, and then putting it out in public where people can criticize it (and you). Humility in that you know that those data and ideas are imperfect and incomplete, and you have to admit openly to the limitations. Too much confidence can blind you to the limitations; too much humility can blind you to the accomplishments.

Getting the balance between confidence and humility right is one of the greatest challenges all developing scientists face, in both doing and writing science. Most of us struggle with confidence—I went through the phase I call "academic adolescence" halfway through my doctoral program, asking, "can I do this?" My advisors were scientists at a level of accomplishment I never imagined I could reach, yet they challenged me to develop and present my own ideas. They pushed me to recognize that I had to do more than just present my results; I had to reach for new knowledge and understanding (remember figure 2.2). They taught me that to do good science, you have to develop intellectual courage and embrace living outside of your comfort zone.

Many, however, are learning not directly from a world-leading scientist but from reading the work of world-leading scientists. That distinction has led to too many papers that basically say "Well-known Professor Genelle found X, and I want to see whether X occurs in my system." I suspect this grows from a sense that "if Prof. Genelle did it, it must be good science, so if I repeat it in a new system, I'll be doing good science, too." When Prof. Genelle did it, it was good science because it was novel. But because she did it, it isn't novel anymore—now it's an old story.

If Prof. Genelle's paper were, for example, "Fungi are more drought tolerant than bacteria in a French grassland," what made it novel was showing that fungi were more drought tolerant than bacteria, not that it was in a French grassland—that's just incidental qualifying information. Showing the same pattern in another system only reinforces her conclusions. For a paper to be publishable in a high-profile journal, it would need a new story. That might be that Prof. Genelle's pattern does not hold in another system, which would pose the question of why they differed. It might analyze the mechanism of enhanced drought tolerance in fungi or evaluate how drought tolerance interacts with other stresses. There would be many ways to take what Prof. Genelle did, figure out what questions her work left on the table or opened up, and ask those. Those would be new questions.

Answering an old question in a new system won't make the science novel. Answering an old question using new technology also won't make the science novel. Even answering an old question in a new system with new technology won't make the science novel. Such work merely fills in the information base. Leading journals look for more than that; they look for papers that provide new knowledge and understanding.

When you develop the courage and ability to ask new questions and take the risks inherent in trying to answer them, you will be prepared to do cutting-edge science. When you push beyond producing information to producing understanding, you will be doing cutting-edge science. Then you will be ready to write the papers major journals are searching for.

19.2. WRITING FOR INTERNATIONAL JOURNALS: TARGET THE RIGHT AUDIENCE

Science isn't complete until it has been published, and the first step in that process is identifying your audience and choosing a journal to submit the paper to. For many scientists (not just those in developing nations), there are competing pressures that can make sorting out story, audience, and outlet difficult. The first pressure is to do research that is practical, solving immediate social problems. The second is to publish in prestigious journals.

The pressure to be relevant can lead to studies that provide information useful to local managers or industries but that may not offer knowledge that would be relevant to a global audience. The pressure to succeed, however, can lead researchers to submit those papers to high-profile journals even when they may not be a good fit. I have seen many of them, and I've rejected the majority, many without even sending them out for review.

Being rejected is painful; no one likes to be told that their work isn't good enough. I've seen authors claim instead that a paper was rejected because the editor discriminated against them or their region. We don't. Rather, the opposite is true—we want to broaden the international base of the research (and reviewing) community. Our problem is that we see many papers that were rigorously done but only offer *information*. In these papers, authors often highlight that what is novel is that it presents the first data on a process in a new region—trace gas emissions, nitrification, and so on. They are usually right, but that very argument is why the paper was rejected.

Any leading journal is likely to reject a paper if all it does is flesh out the information base: it's the first data set on a new region, it demonstrates that a reaction works similarly with a slightly different substitution pattern on a molecule, or that the gene sequence from a new bacterium is only modestly different from that in known bacteria.

This isn't about basic versus applied research. It's about information versus knowledge. First-rate applied research goes beyond presenting a data set—it provides broader insights into the nature of the problem, insights that are useful to people working on related problems and in different areas. For example, a paper on how plowing a soil alters nitrate leaching and nitrous oxide emissions might be valuable for local managers who are trying to maximize crop yield while minimizing groundwater pollution and greenhouse gas emissions. They need the information, and it should be published in an appropriate venue. But unless the paper also offers new insights into the fundamentals of N-cycling or develops a new, transferable management regime, that venue is not likely to be a high-impact basic-science journal—and that will be true regardless of whether the work was done in India or Indiana.

So before you submit, make sure you know a journal's focus and intended audience. Do you want to offer local farmers improved tillage techniques or soil biologists new insights into how bacteria process N? Read a journal's description carefully and analyze the papers it publishes. If you are still unsure, email the editor and ask for advice. Then pick a journal appropriate for your story and intended audience. Don't focus on the journal's status, but on its scope. There will always be a draw toward the journals with the highest impact, but submitting a paper that doesn't fit is a waste of everyone's time and energy. Ultimately, journal prestige means little—the top journals publish some mediocre papers and lower impact journals publish some extraordinary ones. In the modern world of search engines and open-access journals, good papers will be found and cited whereas bad ones will be ignored, regardless of where they are published.

19.3. WRITING THE PAPER

Wrapped up with targeting your audience is figuring out the story. The best general insight I can offer on this appears in the first sections of the book. Be thoughtful, analytical, and critical about your data and ideas. Figure out what is novel in what you did. Remember that there are few data sets so imbued with novelty that they can't be made dull, and few that are so dull that there aren't novel insights that can be drawn from them. It is your job to find the novelty and highlight it. If you've found the novelty, you've done the hard part—nature gives up her secrets grudgingly. We all wrestle with our data sets, trying to figure out their meaning and their story.

It's only after this that specific language skills matter. You must produce a document in which, at an absolute minimum, the right words are used, they are spelled correctly, and the rules of grammar and usage are followed. It is your responsibility as the author to ensure this. Do not submit a manuscript thinking that the reviewers, the editors, or the publisher will fix imperfect English. We won't.[1] It isn't our job, none of us have the time, and the journals don't have the money. Most journals screen papers for language and bounce back those that are not up to an acceptable standard; they won't send them out for review. We have a responsibility not to overwork reviewers by sending them papers that are not ready. The author's job is to make the reader's job easy.

The tool most authors rely on to fix writing problems is their word processor. The spell checker is essential, but it will miss errors like "their" versus "there" and typos that create a real but wrong word, like "from" versus "form." Then there is the grammar checker; this can be useful in catching some errors and it's better than nothing (but not much). As I write, I periodically check on the things it underlines—it catches some real errors, but it makes a lot of mistakes.

Better information is available in any of a number of excellent books and websites. I may be a native English speaker and an experienced writer, but I still have a shelf full of books on grammar and language (see appendix B for a list of my favorites) and I keep a bookmark in my Web browser to the *Oxford English Dictionary*. It is essential to have good references. Countless books have been written for people who are insecure in their knowledge of English. For guides to grammar and usage, shorter is better. You don't need to understand the deepest arcana of English grammar—you need practical, everyday advice. It's no accident that the most battered and coffee-stained book on every writer's shelf is the shortest: Strunk and White, *The Elements of Style*.[2]

The advice most people will give you, however, is not a reference book, but to give your manuscript to an English-speaking colleague to go over before you submit. This can be useful, but I recommend against relying on a friend down the hall as your only language check—at least, not unless they are both a good friend and a good editor. I've sent back too many papers that were edited by friends who hadn't done an adequate job, and I've had some "polite disagreements" with authors who were sure that because their American friend looked over the paper it must be okay. Editing is difficult and time-consuming. Most friends don't have the time, and many don't have the skills, to do a complete and careful word-by-word edit. There are professional services that do this; some are excellent, and they aren't very expensive. Some publishers list editing services on their websites. After spending the equivalent of thousands of dollars to do the research, spending a few hundred more to ensure the final paper is of the highest possible caliber is a small and worthwhile investment. When you need the job done well, use a skilled professional.

1. Actually, many of us do help with language and writing. We know that beginning writers struggle, and most of us want to help. But we usually only do so when it means tidying up and fixing quirks of English, rather than doing a full copy edit. It is also an act of generosity you should not count on. Editors help those who help themselves.

2. If you don't have access to Strunk and White, the original 1918 version by Strunk is available online for free, http://www.bartleby.com/141/.

My suggestion to not rely on an English-speaking colleague changes completely, however, when that colleague is a coauthor. All authors are responsible for a paper's entire content, and that includes the language. Your English-speaking coauthor is responsible for ensuring the language is correct. When reviewers read poorly written papers with coauthors from the United States, Great Britain, or other English-speaking countries, they can be appropriately brutal. They may question whether those authors were actually involved in the paper or whether they merely failed in their responsibility to ensure it was ready to submit. Either way, your coauthor doesn't look good. Unfortunately, as fallout, you may not look good either. If you are collaborating with a native English speaker, make sure he or she will be willing to do the necessary language-editing, and make sure you allow appropriate time to do it.

As a closing story, a colleague of mine questioned whether this book would be useful for scholars for whom English is a second language. She worried that for people who struggle to write grammatical sentences, my focus on storytelling might be overkill. I pointed out that as an editor, when I get a paper where the story is strong but the language weak, I'll send it back to get the language fixed before sending it out for review. If I get a paper where the story is weak I'll just reject it.

So which is more important—getting the grammar or the story down? I'll vote for story every time. You can hire an editor to help with the language. But you can't hire a scientist to help with the science. It's your science and only you can develop the story. Remember, always, that science is not about information; it is about knowledge and understanding. If you can offer understanding, you are most of the way to writing a paper that will be publishable in the world's best journals.

20

Writing for the Public

Put it before them briefly so they will read it, clearly so they will appreciate it, picturesquely so they will remember it, and above all, accurately so they will be guided by its light.

—JOSEPH PULITZER

Many scientists feel that communicating to the public is a completely different beast than communicating to their peers. A much scarier one. We worry that we will come across badly. We worry that we will be misunderstood and our work misrepresented. We worry that our peers will think less of us.

Yet issues such as climate change, disease, emerging technologies, and genetically modified crops pose challenges for society. Policy makers are acting on these issues, based on their own and the public's perception of our science—perceptions that may be mistaken. We need to fix that. We cannot write papers for our peers and assume that the world at large will get the message. We can't even rely on journalists to understand the science and translate it into English (or French, Russian, or Chinese)—they also struggle to understand our work.

People don't understand what we do. They think us experts—authorities who know the answers. From science classes, that's understandable—they are usually about memorizing. But that isn't who we are; we're not experts, we're scholars.

We live where no one knows the answer and the struggle is to figure out the question. This contradiction is at the root of many of our challenges with the public. In the popular imagination we are either Bwaa-hah-hah-ing our way to world domination or we're nerds in white coats. Yet science is wondrous, and scientists are human. We need to do a better job of educating the public, not only about the content of research but about the *nature* of research.

Our avenues for communicating that message have expanded. It used to be that we might speak to a reporter and hope they didn't mangle the story or perhaps write an article for *Scientific American*. Newspapers, however, still exist and publish material sent to them: letters, opinion pieces, and full columns. If you're in New York, getting something published in your local paper may be a challenge, but it's not in smaller communities—papers are looking for content. The Web, however, is where opportunities have exploded, from your own website to blogs to writing Wikipedia entries. We have the opportunity to speak directly to the world. To do that well, you have to be able to tell your story.

The dark, dirty secret of this book is that it is really all about writing for the public. "Science" isn't a different language than "English." It's a dialect, and everything I've said has been about how to write science in a way that makes it closer to common English. Every suggestion has been at the heart of communicating to nonscientists.

The tools I discuss aim at widening your audience by sharpening the story, focusing on the SUCCES elements, and avoiding jargon. To move beyond a scientific audience, you don't have to do any of that *differently*—you just have to do it *more*. And *less*. More story and SUCCES, less complexity and jargon.

People are fascinated by the mysteries of nature. Consider the undying allure of dinosaurs. Or subatomic physics—something no one understands, yet which gets a lot of press. That is driven by clever names (the "God particle") and massive engineering (the Large Hadron Collider). Astronomy and astrophysics also get attention; the images are glorious, and stories of ultimate beginnings resonate deeply in the human psyche. But many obscure topics get picked up as well. For example, I scanned the *New York Times* for 2010 to see what they ran on frogs:

Endangered frogs
How do frogs handle the shock of landing?
Frogs use foams to make nests for their eggs
Tree frogs shake branches to communicate with each other
And of course: A giant frog that "hopped with dinosaurs"

Some of these stories discuss environmental threats, but more are based on "huh, who knew?" Who knew that toads stimulate their foreleg muscles 90 milliseconds before they land from a jump—even if that means firing the muscles *before* they take off?

More than being pure curiosity pieces, these articles show people what science is and how it works. They narrow the gap between science and society and can inspire the next generation of scientists. Having friends and relatives tell you that

Figure 20.1. The public's flow of concern from data to application.

they heard you on the radio or read about your work in the newspaper isn't the end of the world either.

The other reason the public pays attention to us is because scientific issues dominate modern existence: technology, medicine, and environment. With such stories people are going to ask "how does this affect *me*?" As scientists, we work on these topics because of their fascination—I didn't become a soil microbiologist to save the world but because bugs in dirt are cool. The work affects us by offering new insights into nature. For the public, though, that won't cut it. They are looking for application (figure 20.1): what can we *do* with that understanding?

Figure 20.1 illustrates one reason why people react differently to stories about medicine and global warming. With medicine, everyone perceives application as making their lives better. With global warming, some people fear that addressing the threat may make their lives worse. The pushback on application cascades all the way back to the data.

20.1. WRITING STORIES FOR THE PUBLIC

The way the public perceives science suggests several rules for writing our stories for them.

1. Story: a simple story is critical.
2. Focus: it must target either the "who knew?" elements or the application—what does this mean for them or for society? It helps to talk about the people doing the work. Scientists are trained to downplay the human dimensions, but for nonscientists, people are more engaging characters than chemicals or concepts.
3. Story structure: the audience is impatient and itching to know why this story will be interesting. That calls for a front-loaded structure, most likely LDR, to quickly engage the SUCCES elements.
4. Language: keep it simple. Assume that you need to define every technical term, or better yet, find a way to avoid them. In writing for the public, a technical term will likely feel like jargon. Importantly, never introduce a technical term in the topic position of a sentence. Finally, don't be patronizing; assume your audience is intelligent but ignorant. You are not "dumbing down" the story, you are highlighting its simple core and telling it in their language. Your job is to interest and educate your audience.
5. Scientific method: The public thinks that science proceeds through "Eureka!" breakthroughs made by brilliant men and women and expects

stories to present those breakthroughs. But that isn't how science works. It advances through incremental advances by many people struggling to collect difficult data and make sense out of the puzzles that emerge. We are about process, not product—we don't expect to achieve the ultimate breakthrough, though we are driven by the search. The best stories for the public integrate our joy in puzzling out nature with their focus on results and application.

20.2. THE MESSAGE BOX: A TOOL FOR FRAMING STORY

A useful tool to help integrate these ideas and frame a story accessible to the public is the message box (figure 20.2), a concept described well by Nancy Baron.[1] The message box is a simple graphical tool for laying out a story's essential elements. It will help you sort through and define the SUCCES elements.

A and B. Issue and Audience. Start by identifying your overall issue and audience. Keep your eye on them as you develop the story. These aren't specific elements but overarching themes to weave through the story.

1. Problem: what is your specific problem? This defines your opening and your challenge. In a story for the public, these should be conflated into a clear lead.

2. So what? Moving clockwise around the box, we come to "So what?" Why should readers care about your problem? Your "so what?" must be appropriate for your audience; it will be different for your colleagues and relatives. If you are writing an LDR-structured story for the public, your "so what?" should be integrated into the lead.

3. Solution: having defined the problem, offer your solution. This will comprise the body of the story: the A in OCAR, the D in ABDCE or LDR.

4. Benefits: finally, what are the benefits of your solution? What would your audience get from it? This should be your resolution.

20.3. HOW TO TURN YOUR SCIENCE INTO THEIR STORY

As an example of using the message box to craft your science into a story for the public, I pose the following scenario. You're a researcher working on the networks of genes that regulate cell growth; these form cascades in which there is a series of triggering events, with one gene regulating the next, which regulates the next, ultimately controlling cell growth. When this regulatory cascade fails, cells switch from normal growth to an out-of-control state that converts them to cancer cells.

---

1. N. Baron, *Escaping the Ivory Tower: A Guide to Making Your Science Matter* (Island Press, 2010).

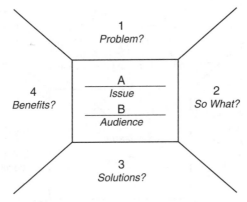

Figure 20.2. The message box.

The model system you're working in is zebrafish, because it is convenient: they are easy to grow, the gene cascades are similar to those in humans, and the genetics are tractable.

This is fascinating basic research about the processes that regulate growth and life. It raises interesting questions about complex feedbacks and sophisticated regulatory mechanisms. But your aunt, or a reporter, will ask whether you're going to cure cancer.

You know your work isn't going to immediately cure cancer. At best it will provide a small insight into how cells work and how they break. Coupled with the work of hundreds of other researchers, that insight may eventually contribute to creating a treatment that might prevent or cure some cancers. You may feel that saying that your work is going to "cure cancer" is such an outrageous stretch that you don't want to even let those words slip from between your clenched teeth.

But you have to answer your aunt's question in terms that make sense to her. That explanation must start with her concerns and her schema—curing cancer. If you can't match her there, she'll wander off, impressed with how smart you are but completely baffled about what you do, wondering why you didn't go to medical school and become a doctor. A reporter might wonder why tax dollars are going to support your Ivory Tower hobby.

To develop a story about your research that your aunt or that reporter will understand, start by building a message box (figure 20.3).

A. Issue: curing cancer.
B. Audience: your aunt specifically, and the public more generally.
1. Problem: we don't understand the complex cascade of genetic interactions that cause normal cells to "go bad" and start growing out of control.
2. So what? The switch from normal to out-of-control growth is the critical step in developing cancer.
3. Solutions: we need to understand the genetic interactions that cause this switch so we can prevent or reverse it. To study these interactions,

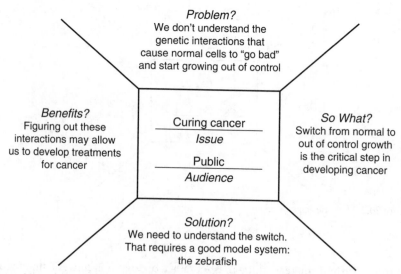

Figure 20.3. A message box for studying gene cascades in zebrafish.

we need a model system. Zebrafish turn out to be a surprisingly good system because they are easy to grow, their genetics are easy to work on, and they function similarly to humans.

4. Benefits: figuring out the genetic interactions that cause normal cells to become cancer cells may allow us to develop treatments that would prevent or treat cancer.

Now we can take those points and write a story, using simple language.

Example 20.1

Gene Cascades in Zebrafish

Cancer is what happens when normal cells start growing and dividing out of control. If we want to prevent cancer, we need to know what causes that switch—why do "good cells go bad?" My research targets that question—I study how genes interact with each other to keep cells working and growing at the "right rate," and how those interactions break down, turning normal cells into cancer cells. I work on zebrafish because their genes behave similarly to those of people—and you can't grow people in an aquarium. If we learn what causes growth regulation to break down, we may be able to prevent or reverse it. So, yes, I hope that my work will ultimately contribute to curing cancer.

This is accurate, honest, and doesn't make any grandiose claims. It also frames the research in a way that your aunt can understand—both why it's intellectually interesting and how it might contribute to curing cancer. This might be enough of an answer by itself, or it might constitute the opening for a larger story. Having laid the framework, you could easily take the next step and introduce more complex concepts, like gene cascades.

This short example works in part because it has a clean OCAR structure and plays effectively on SUCCES elements. It is simple, focusing on the breakdown in gene regulation of growth. It is emotional, addressing a question your audience cares about. It is concrete and credible in terms of describing the work and how it fits into the overall scheme of the search for a cure for cancer. Finally, it uses simple language. It might even convince your aunt that what you're doing is a reasonable alternative to medical school.

20.4. USING THE MESSAGE BOX MORE GENERALLY

The message box can be a powerful tool for writing for your peers as well. Our questions and concerns as scientists may be different than the public's, but when we read we have the same questions: what is the problem, what is your solution, why should I care?

I filled out a message box for my project evaluating microbial activity in arctic tundra soils during the winter (figure 20.4). On the surface, the work may seem pretty abstruse—why would we care about microbial activity in frozen soils of the tundra? A message box helps identify the main points to weave into the story. Example 20.2 is condensed from the opening of a proposal I wrote to support this work.

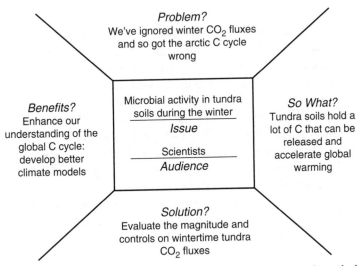

Figure 20.4. A message box for studying microbial activity in arctic tundra soils during the winter.

Example 20.2

Microbes in frozen soil?

Arctic ecologists have traditionally focused on the growing season, the short period between snowmelt and snowfall. As researchers packed up in August to head home for the academic year, there was a sense of "last one out, turn out the lights." Yet, the Arctic is dominated by the 9-month-long winter. For many years, it was assumed that plants and microbes were frozen into dormancy for the winter. Soil microbes, however, continue to respire through the winter. While the rate may be slow, Arctic winters are so long that this respiration constitutes a large fraction of annual carbon budgets, a fraction that could increase with climate warming, which is greatest during winter.

Because Arctic tundra contains half of the global soil organic C—more than twice the C that is in the atmosphere as CO_2—an increase in the rate of wintertime microbial respiration could increase atmospheric CO_2 and intensify global warming. Developing a better understanding of the magnitude and controls of winter microbial respiration is therefore critical to building models that can project the future climate.

This version condenses the actual text (I shortened the paragraphs by deleting technical material and literature citations), but the flow of the argument, the story, is identical. I even included the "last one out, turn out the lights" comment in the proposal. Using the message box helped structure the story of the proposal, but maybe more important, it prepared me for when a reporter called about a piece he was doing for the *New York Times*.[2]

20.5. DEALING WITH POLICY MAKERS

Communicating clearly may be enough if you are trying to explain what you do to friends, relatives, or reporters. It is not sufficient, however, if you are trying to work with policy makers and managers to influence decisions and actions. Many scientists believe that if these people understood the science, they would make appropriate decisions. But suggest that to an elected official or an agency manager and you can expect either a snort of derision or hysterical laughter. Their decisions are influenced as much by political as by natural reality.

If you want to influence policy, learning to speak effectively to decision makers is a start. Part of this is language: congressional staffers are usually 22-year-olds with B.A.'s in political science, not Ph.D.'s in chemistry. More of the communication is story structure. You've got their ear for five minutes in a hallway—an audience impatient enough to make a *Nature* editor look languid. A staffer may hear what a scientist says as "Blah, blah, carbon, blah, blah, respiration, blah, blah.

2. Paul Voosen, "Even When Frozen, Soils Get Busy Emitting CO_2," *New York Times*, November 29, 2010.

If we wait until the permafrost melts to do something about global warming, that will be bad." Those last words may be the ones we thought about the least, but they are the ones they listen to the most. That is where they listening for your wrap-up message—what you are asking them to do.

Beyond language, however, are culture and motivation, and they are alien to those of academe. A striking illustration of this for me was when I was an assistant professor at the University of Alaska. Shortly after I started, the *Exxon Valdez* crashed into Bligh Reef, and as with most Alaskan scientists, I got swept up in the damage assessment. One of the managers told us "this isn't about science, it's about damage assessment, but you have to use the best available scientific methods." He meant sophisticated analytic tools; it didn't seem to occur to him that it should include doing good science, which I consider *the* essential scientific method. Sophisticated but misapplied measurements have less value to me than do simple, but smart ones. The lawyers described what we were doing as "you're not collecting data, you're collecting evidence." Managers and politicians face pressures beyond the science. They don't even see science in the same terms we do. Effecting change means working in their world, not expecting them to work in ours. To be effective in the policy arena requires understanding and political skills that go beyond the communication tools that are the focus of this book.

EXERCISES

20.1. Fill out a message box for your short article, with the intended audience as scientists.

20.2. Fill out a message box for your short article, with the intended audience as the public.

20.3. Rewrite your short article for the public.

21

Resolution

As a scientist, you are a professional writer.

When I was a student, it always bugged us that they called graduation ceremonies "commencement." We knew that we were celebrating finishing, earning our degrees after years of hard work, and that we were also mourning finishing, ending years of comradeship. The idea that the ceremony was a beginning was absurd, and to give it the pompous name of "commencement" only made it worse.

We were wrong, as I realized the day I finished my Ph.D. After floating out of the library having filed my dissertation, I looked at the receipt they had given me and came crashing down. All my life I had been a student; my entire self-image was built around being a student. It's who and what I was. Yet that silly slip of paper said I wasn't one anymore. So who was I?

No longer at the pinnacle of the student world, I was now at the bottom of the professional. The most junior, inexperienced, and unproven Ph.D. in a world of Ph.D.s. The degree I'd struggled for meant nothing; it was just an entry ticket to this new arena, where I had to start proving myself all over again. Indeed, all graduations are commencements.

Now we come to the resolution of this book, and it is both graduation and commencement. Here I tell you that you have learned new tools and skills and congratulate you on your accomplishments. But this, too, is a new beginning.

21.1. LOOKING BACK: THE LESSONS

Reading this book may have expanded your perspectives on writing and communicating as a part of what you do. I hope the philosophy of storytelling will make you better at highlighting what is important in your work, and so make you a better scientist. I hope as well that it will make you better at communicating that work, whether it be for the readers of *Cell* or for your grandmother. I focus on writing, but the messages are about communication generally. After all, Homer didn't write the *Odyssey*—he recited it. Only centuries later did someone capture it in the written word. But Homer understood OCAR and SUCCES.

The most important message I have tried to emphasize is the one I started with: *as a scientist, you are a professional writer.* Applying the tools of the writer will make both your writing and your science stronger. Those tools will solve problems from whole papers and proposals down to individual sentences. Within each chapter I offered guidelines for dealing with specific issues, but they all grow from principles of stickiness and story structure: SUCCES and OCAR. Remember to target your audience. Remember to focus on the critical story elements. Remember that science is about knowledge and understanding, not just data.

All the tools I offer will not enable you to turn a lump of coal into a diamond. Writing skills cannot replace other scientific skills—if the research is fundamentally weak, you can't use clever writing to strengthen it. But writing skills will help you take a rough and flawed gem of a data set, identify its valuable core, facet it, and polish it to play up its assets.

Working through the chapters and doing the exercises should have expanded your tool box, but that won't make you a writer. Now you have to take those tools and work with them to develop expertise, deepening your insights and abilities. As you do that, your skills should mature to the point that you produce papers with maximum power and reach and proposals with maximum fundability.

21.2. LOOKING FORWARD: BECOMING A SUCCESSFUL WRITER

When you opened this book, you were probably a scientist who writes, a scientist who wants to write, or perhaps a scientist who has to write. I hope this book has set you on the road to becoming a scientist-writer. Learning to write, however, is a road without end. I was proud of the first proposal I got funded, until I read it later and wondered what drugs the review committee had been taking. I thought I knew enough about writing two years ago to write a book on writing.

I can't tell you how much I have learned in the last few years, and how much better my writing has become. The tools I offer are just a starting point for becoming a successful writer and developing a successful career in science.

21.2.1. Looking Forward: Becoming a Successful Scientist

As you start your career, some will give you advice on survival strategies. I think that is a mistake—I prefer to focus on success strategies. If you clear the bar of success, you will clear the bar of survival without noticing it was there.

Survival strategies usually distill down to "publish or perish." Unfortunately, some colleagues interpret this as arguing that quantity is more important than quality. They argue that you survive by publishing and succeed by publishing a lot. They recognize that administrators rarely work in our immediate field and may not know whether our work is important, but they know that those administrators can look at our CVs and count our publications.

"Publish or perish" may be the basis for survival, but it is *not* the basis for success. As I argued in chapter 1, you don't succeed by getting papers published but by getting them cited. Ultimately academic success is grounded in a piece of advice my brother gave me once when I was frustrated with university politics. He said simply "remember who your real peers are." That meant soil and ecosystem scientists around the world, the people who read my papers. Your real peers will see through flash, splash, and numbers to identify papers that contribute. They'll cite those.

Impress your real peers, and you will impress your tenure committee and your dean. Your dean can see your citation history, and can read your record of invitations to present at conferences and serve on committees and editorial boards. Those speak loudly about your standing in the field. We also rely on peers to send us students and postdocs—the people who do the work and write the papers—so quality begets quantity.

Most important, in the U.S. system the decisive part of a promotion package is letters from your peers. When I write letters assessing people for promotion, I usually don't know exactly how many papers they've published, nor do I particularly care. I do know whether they've been creative. I do know whether their papers have influenced my thinking. I do know how those papers have moved the field. I sometimes even know whether they've done work so clever that its value has been missed by the crowd. I respect colleagues whose papers are thoughtful and insightful, and I've gone to bat for them. I've fought for colleagues who don't write lots of papers, but who write ones worth reading and citing.

You may survive by publishing a lot of papers, but you will only succeed by writing good ones—papers that are clearly structured and tell a compelling story. Quality ultimately trumps quantity, and it will stand out in a crowded scientific universe. After all, Einstein's face isn't on T-shirts and advertisements around the world because he published a lot.

Good luck and good writing.

My Answers to Revision Exercises

5.3A: The problem with this example is that the opening sentence doesn't give direction. This is trying to be a pawn push, introducing the idea that all chemical reactions are temperature-sensitive and preparing for the argument that this should be true in soil. However, an effective opening must introduce the story's central character and major issue; that is not "all chemical reactions" but soil respiration. So soil respiration should appear in the opening sentence.

Respiration in soils, surprisingly, doesn't always increase with temperature as predicted by transition state theory and the Arrhenius equation. Some studies have shown no respiration response to increasing temperature, while a few have even reported a negative response.

5.3B: This is trying to capture a very wide audience with a broad statement about the importance of chemotherapy, but it is an example of bad misdirection. Everyone knows that chemotherapy is a common treatment for cancer, so a reader would just skip over that and get caught by "development of new targeted-delivery systems." That is an exciting and novel idea that would lead readers to assume the paper is going to tell us something about such systems. In fact, that was just an add-on to argue that chemotherapy is going to become even more important than it already is.

The real story, however, is not about drug delivery systems but overcoming resistance to treatment. This opening needs to get straight to that idea and cut out distracting bells and whistles. Delete the first sentence and then adapt the second as the opening.

A common constraint to effective cancer chemotherapy is that patients may be resistant to the treatments. Such resistance is often closely associated with the activity of the enzyme γ-glutamyl transpeptidase (GGT), which acts to increase intracellular concentrations of glutathione and thereby block the apoptotic cascade in tumor cells. Inhibiting GGT before chemotherapy would therefore reduce tumor cell resistance and increase treatment effectiveness.

11.3A

It had been thought that the lack of jet contrails over the United States caused the increase in the diurnal temperature range (DTR) during the three-day grounding of aircraft over the United States during the period of 11–14 September 2001. Variations in high cloud cover, including contrails and contrail-induced cirrus clouds, contribute weakly to the changes in the diurnal temperature range, which is governed primarily by lower altitude clouds, winds, and humidity. While missing contrails may have affected the DTR, their impact is probably too small to detect with statistical significance.

This poses the question in the first sentence, rather than stating the conclusion. For that to work, the remainder of the paragraph would need to present some of the results and build the argument to support the conclusion. This worked as a TS-D paragraph, because it is apparent that the core results are presented elsewhere and this just presents their essence. I think the author's point-first structure worked better than a point-last.

11.3B

Great Plains mammoths apparently did not routinely migrate long distances, such as between northern Colorado and southern High Plains sites that are separated by about 600 km. Mammoth samples from Clovis sites in the Dent site had different $^{87}Sr/^{86}Sr$ ratios than those at Blackwater Draw and Miami, indicating they come from distinct populations.

To me, this paragraph feels weaker. The original poses a question and proposes a case study approach to answering it. To readers, the constraints on the approach are clear. The TS-D version starts with an argument that makes the constraints less apparent. TS-D is a weak structure for developing a complex argument—it needs a simple topic sentence. The OCAR structure allowed the author to develop both the simple core story—mammoths did not migrate long distances—and its limitations. The TS-D version is shorter, however, and if space were at a premium that might be more important than the nuance allowed by OCAR.

12.3A

1. Viruses are the most abundant biological entities in the sea, yet were not studied until 1989.
2. It wasn't until 1989 that it was discovered that the most abundant biological entities in the sea are viruses.
3. It wasn't until 1989 that it was discovered that viruses are the most abundant biological entities in the sea.

12.3B

1. Benzene contamination of groundwater is linked to elevated cancer levels.
2. Groundwater contaminated with benzene can cause cancer.
3. Cancer levels are higher in areas where groundwater is contaminated with benzene.

12.4A: The story in this sentence is that we don't know the crystalline structure of kryptonite, so the critical word is "unclear." That word should therefore be the stress, instead of being buried in the middle of the sentence. So move it to the end:

Due to uncertainties resulting from interferences in the X-ray microanalysis, the crystalline nature of kryptonite remains unclear.

12.4B: The actor in this story is "drought," so it should be the topic. If we can't delete the initial clause, we can move it into the middle of the sentence.

Drought reduces soil microbial activity by reducing diffusion and increasing physiological stress, causing a build-up of biodegradable C that is rapidly respired upon rewetting.

13.3A: There is no stress–topic linkage between these sentences. So make Fe the topic of the second sentence.

Studies comparing iron-resistant and sensitive cell lines confirmed that protein X17 is denatured in the presence of Fe. When cellular Fe concentrations decrease, however, protein X17 reverts to its native form.

13.3B: The critical argument in the first paragraph is that the two promoters are synergistic: together they are more effective than either alone. The second paragraph adds a new twist to explaining the synergy, but it starts by introducing a new character. The connection to the previous paragraph is unclear. Reach back and grab the idea of synergy.

The synergy between TREE2 and STEM3 that allows full transcription of *tryb* appears dependent on the presence of LEA. When LEA was absent, even with both promoters intact, transcription rates were only 50% of control levels . . .

14.3A: This sentence has only two verbs in it: *increase* (opportunities) and *alter* (population dynamics). But what is the critical action? It is "ecological interactions." What does that mean? It's a euphemism for a nematode eating a bacterium.

That is dramatic, so make that action the verb; if *eat* seems too intense, you could use *consume*.

> Increased mobility of predatory nematodes in soil would allow them to consume more bacteria and so alter bacterial population dynamics.

14.3B: Two things to note here: all the actions are expressed in nominalizations, whereas the verbs are fuzzy and carry no sense of the action.

Action	Expressed as	Verb associated with the nominalization
Challenge	Challenges	Present
Remediate	remediation	Present
costs money	financial costs	Invoke
threatens health	health risks	Present
Expose	exposure	Face

The following version uses the same core words to express the ideas of costs and threats but expresses them in verbs.

> Polyaromatic hydrocarbons and polychlorinated biphenyls are challenging to remediate: it costs a lot of money and it threatens the health of workers who are exposed to the compounds.

14.3C: This includes most of the problems I've discussed.

was demonstrated	Fuzzy and passive
extraction	Nominalization
an enhancement	Fuzzy and nominalized
an extraction	Nominalization

> Extracting soils with NH_4Cl instead of K_2SO_4 enhanced Al recovery.

Here, I applied some of the rules from chapter 12 and restructured the sentence to unbury the stress as well as to energize the verbs! And no, that isn't cheating. You are always going for the best possible sentence.

That wouldn't be: "Extracting soils with NH_4Cl enhanced Al recovery relative to extracting with K_2SO_4."

15.3A. This sentence suffers badly from prepositional phrases and nominalizations. It also suffers from a buried stress: the important message is that studies haven't been done.

> Animals' abilities to solve problems have been undervalued because scientifically reliable studies have not been done.

15.3B: A lot of heavy words.

> Rats kept under varying environmental conditions had better cognitive ability that the control group, which had been kept under constant conditions.

16.2.A

> Polyaromatic hydrocarbons and polychlorinated biphenyls are challenging to remediate: it is expensive and dangerous for clean-up workers.

This is shorter, and I think it is stronger. I shortened it by condensing "costs a lot of money" and "threatens the health" into nominalizations—"expensive" and "dangerous"—and by switching an Anglo-Saxon word (*threaten*) for a French one. I broke the rules, but I collapsed short phrases into well-understood words. I think the trade-off worked. I also replaced "workers who are exposed to the compounds" with "clean-up workers." These chemicals are toxic to everyone, but since we're talking about remediating, the workers of particular concern are those doing the clean-up.

16.2B

> *Chla* and *Chlb* transcript abundance showed similar patterns in plants at different developmental stages.

I condensed this by deleting metadiscourse. We don't need to know about the comparison or the observation, just about the transcripts.

16.2C

> Inherent resistance is an evolved response to living in constantly harsh environments. Resistant plants don't induce a physiological response to stress; rather, they have traits such as high root biomass, extensive chemical defenses, and low growth rates.

The biggest problem with the paragraph is that "inherent resistance" is repeated three times. There are several other expressions that are cumbersome—prepositional phrases such as "environments that are constantly harsh" and "characterized by traits." By cutting these out, I was able to collapse two sentences into one.

BOOKS

General Writing

Style: Toward Clarity and Grace, Joseph Williams (University of Chicago Press); *Style: Ten Lessons in Clarity and Grace* or *Style: The Basics of Clarity and Grace* (both from Pearson Longman). The different versions of *Style* are, in my opinion, the best books on writing English that exist. I prefer the original *Style: Toward Clarity and Grace* because it's more analytical and less of a pure textbook.

The Elements of Style, William Strunk Jr. and E. B. White (Longman). The essential reference.

Writing Tools, Roy Peter Clark (Little, Brown) and *The Glamour of Grammar*, Roy Peter Clark (Little, Brown). These are both insightful, useful, and entertaining. They range from basic to very advanced insights into writing. Clark is a leading teacher of journalism and it shows.

Approaches to Writing

Bird by Bird, Anne Lamott (Anchor Books). A wonderful book by a talented fiction writer. Many of the insights transfer to science writing.

On Writing Well, William Zinsser (Collins). The classic guide for journalism and nonfiction. Science is supposed to be nonfiction!

Made to Stick, Chip and Dan Heath (Random House). A brilliant guide to communication strategy. So much so that I spent all of chapter 3 to reprise it.

Communicating Science

Eloquent Science, David M. Schultz (American Meteorological Society).

Essentials of Writing Biomedical Research Papers, Mimi Zeiger (McGraw-
Hill).

These are both extensive and technical guides to writing science, each targeted
at a specific disciplinary area. The information in them is detailed and excellent.

WEBSITES

Oxford English Dictionary: http://www.oed.com/. This is the essential
language resource in English.
Merriam-Webster: http://www.merriam-webster.com/. It includes both a
good dictionary and a thesaurus.

2 – 3 – 1 rule of emphasis 116, 148, 156

Abbreviations 149
ABDCE story structure 27–30, 50
  in proposals 56
Abstractions 22–23
Acronyms 149
Action 96
  in ABDCE structure 27–28
  describing action 67–81
  in OCAR structure 27
  in a sentence 112, 116, 134
Active voice 134
  vs. passive, debate 137
Actor, in a sentence 116
  in active vs. passive voice 134
  hiding 135
Adjective 163–166
  nominalizations 142
Administrators 206
Adverb 138, 163–164
Anglo-Saxon 151-2
Appendices, in a paper 74
Applied vs. basic research 192, 197
Archives, data 74
Arc, story 95–100, 125–128
Aristotle 47
Audience 4, 21
  broad vs. narrow 35, 147
  patient vs. impatient 27–31, 60, 183
  targeting a specific 21, 40–45, 191

Author (*also see Writer*)
  inexperienced 193
  responsibility of 193
Azam, Farooq 41, 44

Background material
  in ABDCE story structure 28
  in IMRaD 32
  in OCAR 50, 56
Baron, Nancy 198
Beginning
  as part of a story 26
  as place to introduce new ideas 149
  as "power position" 35, 97
Bizzwidget 54
Bureaucrats 133, 135
"But, yes" approach to limitations 180, 187

California Environmental Protection Agency 41
Career
  role of writing in success 5, 13, 206
  success vs. survival 206
Chandler, Raymond 8
Challenge, in OCAR structure 27, 58–65
Characters
  in a story 9, 25, 28, 197
  "listening to" 11, 19, 72, 137
  scientific concepts as 9, 36–38, 52, 98–101
  as topics of a sentence 112–113

Chess 47
Chicago Manual of Style 149
Churchill, Winston 113, 143
Circle, story as,
    closing (resolution) 28, 81, 91
    within a paragraph 127
    vs. spiral 29
Citation
    as measure of impact 3, 5, 16, 206
    in establishing credibility 23
    in a literature review 56
    role in success 3, 45, 91
Clarity
    clear thinking vs. clear writing
        4, 7, 91
    in SCFL editing process 175
Clark, Roy Peter 23, 116, 120,
        160, 166
Clause
    in hierarchical structure 96, 120
    linking 141
    main 120
    opening 119, 185
    parenthetical 148
    qualifying 117
    stressed 113, 148, 156
    subordinate 120, 148
Climax, in ABDCE 28–29, 99
Clinton, Bill 17, 22
Coherence
    thematic 101, 124, 128
    and story arcs 102
Compartmentalizing thoughts 97
Compound nouns 153–154
Communication
    cross cultural 8
    miscommunication 143
    with policy makers 202
Conclusion(s)
    constraining 180, 187
    foreshadowing 35
    how general should they be 33
    in a paragraph 105, 107
    question as a conclusion 86
    section in a paper 27, 83–92
    "telegraphing" 60
    undermining 91

Concrete, being
    common language 152
    element of SUCCES formula
        22, 133
    importance of 63, 89, 139
    names 143
Condensing 158–172
    arguments 17, 56
    and clarity 110, 170
    data 72–73
Confidence 7, 70, 190
Constraining
    audience 5, 45
    conclusions 65, 90, 180–188
Courage 140, 190
Credible in SUCCES formula 23,
        133, 201
Crossley, D.A. 32
Curiosity 21, 24, 41, 51, 55
Curricula in science classes 20
Curse of Knowledge 22, 110

Darwin, Charles 8, 18
Data
    choosing data to present 72
    collection as research goal
        59, 64
    data dump 54, 56
    "data not shown" 133
    finding novelty in 21, 192
    imposing story on 31
    vs. inference and interpretation
        70–72
    statistical results as data 76
    synthesizing into understanding
        8–12, 88, 192, 197
Deleting 161–167
    to focus on key points 39, 118
Development, as story section
    in a story arc 96
    in ABDCE structure 28
    in LD structure 29, 104
    in LDR structure 30
    in OCAR structure 50
    in TS-D paragraph 104
Dietrich, William (Bill) 12,
        37, 47

Direction
    changing 35, 91, 101
    defining 68, 110, 175
    misdirection 38
Disciplines, academic 18, 32, 40, 42–43,
        59, 70
Discussion (section of a paper) 8, 79, 96
    addressing limitations in 184–187
    combining Results and Discussion
        70–72
    in IMRaD 32–33
    as part of the Action (A) in OCAR
        67–68
Dogma 21, 56, 169
Drafts 5–7, 39, 158, 160, 174
"Dumbing down" 17, 178, 197

Editor(s) 31, 70, 191, 193
Emotion, in SUCCES formula 24, 50
Emotional weight. 151
Empty amplifiers 164
Ending(s)
    in ABDCE structure 28
    as power position 83, 97, 113
English
    history of 151
    as a second language 189–194
Expert 20–22, 147, *also see schema*
    vs. scholar 196

Fields of science *see disciplines*
Firestone, Mary (Ph.D. advisor) 12,
        178, 190
Flag word 56, 84, 85, 99, 100, 102
Flow, creating 100, 124–132, 172, 175
Formatting, of a page 159
Franklin, Rosalind 10–11
French 151
Friend, using to review a manuscript
        193
Front-loaded story 28–32, 60, 92,
        134, 197

Graduation ceremony 204
Grafton, Sue 28
Grammar 112, 134, 194
    checker 193

H-factor 3
Headers of sections 62, 69, 100
Heath, Chip and Dan 16
Homer 28, 205
Honesty 79, 139, 180
Hourglass, structure of a paper 33, 45,
        50, 89
Humility 91, 190
Hypothesis 32, 58, 65
    falsifiability 58, 138
    fuzzy hypotheses 138

Ignorance 21, 56, 147, 197
Impact Factor 3
IMRaD structure 32
Inference, vs. data 70–72
Information
    processed 97
    synthesizing into knowledge 11, 24,
        56, 59, 191
    sequence of new vs. old 113
Interpretation vs. results 21, 70, 78
Introduction
    as section of story 8, 50, 96
    in proposals 32
    in IMRaD 32–33
    bad introductions 53–54
    vs. literature review 55–56
    in dealing with limitations 181
Inverted pyramid 29
Ivory Tower 10, 198

Janus (function) 125–127
Jargon
    abstractions as 23
    acronyms 149
    avoiding 147, 196
    definition of 146–147
    nominalizations as 143
Journal
    generalist vs. specialist 31, 60,
        62, 115
    impact 3, 16, 189–190
    international 190–191
    open-access 192
    requiring data archival 74
    scope 27, 45, 192

Journalists
    communicating with 8, 195
    newspaper vs. magazine 30
    objectivity 9

King, Stephen 134, 160, 163
Knowledge
    curse of 22, 110
    gap 21, 41, 50, 54, 86, 181
    synthesizing from data 11, 197
    schemas 21, 39
    vs. information 24, 58, 63, 65, 88,
        192, 205
Kolbert, Elizabeth 8

Ladder of Abstraction 23
Lamott, Anne 5, 7, 9, 29, 95, 137
Language
    colorful 41, 137
    common 147
    fuzzy 54, 92, 137–139, 152
    literary 85, 145
    and objectivity 137
    in SCFL acronym 175, 177
    "science" as a 196
    simple vs. complex 20, 170, 197
    written vs. spoken 146
Latin 58, 151–153, 172
LD story structure 29
    describing methods and results
        68, 74
    in paragraphs 104–106, 126
    for sentences 120
LDR story structure 30
    in discussion section 79–81
    for generalist audiences 47
    in paragraphs 107–109
    in proposals 32, 92
    writing for the public 197
Lead, element of a story 29, 198
    false 38
    in science papers 31, 62
    in proposals 92
Leaders (in their fields) 4, 140, 190
Learning 20, 105
Limitations, addressing 91, 180–188
Lincoln, Abraham 17

List (vs. story) 126
Literature review 55

Magazines 30
Materials & Methods 32, 68
    as story arc 96
    using first person 136–137
    addressing limitations 182
Mentorship 140, 190
Message box 198- 201
Metadiscourse 166
Methods, describing 68
Middle
    as section of a story 26
    of a sentence 115–116, 149,
        also see 2-3-1 rule
    of a paragraph 184
Misdirection 38
Mistakes
    learning from 39
    "were made" 135
    by grammar checkers 193
Modifiers 163, *see adjectives, adverbs*
    good 165
    in prepositional phrase 153
    redundant 161
Montgomery, Scott 4
"Murder your darlings" 74

National Academy of Sciences 12, 18
National Institutes of Health (NIH) 47
National Science Foundation (NSF) 7,
        41, 59, 94, 155
    grant proposal guide 158
    targeting different programs 43, 59
Nature (journal) 31, 47, 53
New York Times 196, 202
Newspaper
    story structure 29–30
    writing for 196
Nominalization 140–143, 150, 153
    adjective 142
Noun
    clusters 155
    compound 153–154
    noun train 155, 177
    instead of adjectives 163

Novel(ists) 95, 124
Novelty 12, 21–24, 186, 190–192
Novice 22, 69, 145, 149

Object
    of a sentence 112, 134
    of a prepositional phrase 153
Objectives (of research) 54, 58–60, 65
Objectivity 9, 137
Obvious, as a target for deletion
    162–163
OCAR 27–30, 189, 201, 205
    applying to science 31–33
    background material 50
    in Discussion section 79
    and paragraphs 107
    and sentences 112, 120, 134
    telegraphed 60
Old English 151
Opening 35–47
    in defining an audience 40, 42,
        44, 55
    in OCAR structure 27
    of a paragraph, 105, 107, 138
    role in developing flow 125
    in scientific papers 32–33, 50
    of a sentence 112, see *topic*
    in story arc 95–96, 125
    two-step 42
Oxford English Dictionary (OED)
    152, 193

Page limits 69, 94, 152, 158
Paragraph(s) 104–110
    incoherent 101, 109
    LD 106, 124, 184
    LDR 107
    linking 129–132
    OCAR 108
    point-first 105
    point last 107
    in story hierarchy 96
    Topic sentence (TS-D) 104
Parallelism 143, 177
Passive Voice
    definition of 134
    limitations of 134, 141

uses of 129,135–138, 156
    objectivity 137, 167
Pawn push 46–48, 165, 176
Perfectionism 6, 91
Perspective, controlling 134–135
Philosophy of science 31, 58, 78
Platitude 26, 46
Plot
    imposing 72
    in story development 9
    twists 91
Policy, influencing 10, 195, 202
Popper, Karl 58
Positioning statement 37, 46, *see pawn
    push*
Power position 35, 47, 83, 91, 97, 113
Prepositional phrases 153–155
Principles
    adapt to different media and
        audiences 44
    all tools in English have value
        137, 154
    author's job to make the reader's job
        easy 5, 59, 150
    make the paragraph the unit of
        composition 104
    OCAR as a principle 32, 112
    others should be able to repeat
        work 68
    must distinguish results from
        interpretation 70, 78
    show, don't tell 133
    SUCCES as 205
    vs. rules 32, 127, 131
Problem, scientific
    defining 18, 26, 35–37, 50, 198
    failing to define 53–54
    solving 79, 83
Problem, writng
    creating new 174, 178
    diagnosing 129
Professionalism 3–6, 9, 14, 44, 146, 178
Proposal(s)
    addressing limitations 183
    hypotheses 138
    page limits 152, 158
    resolutions in 92

Proposal(s)(*Cont.*)
   reviewers 24, 31, 40–43
   structure of 27, 32, 56
   success/funding 5, 24, 139
Public, the 10, 143, 195–203
"Publish or Perish" 3, 206
Pulitzer, Joseph 174, 195

Queen launch 46, 165
Question
   importance of 21, 24, 32, 51, 58–59
   specific 33, 50
   boring 21, 32, 139

Reader(s)
   expectations 5, 11, 30, 36, 45, 129
   schemas 20, 39, 63, 113, 143, 147
   patience 27, 31
Redundancy 161
Rejection 158, 191
Research, basic vs. applied 192
Resolution 27–28, 33, 83–94
   in LDR structure 30
   in a paragraph 107, 129
   in proposals 92
   question based 86
   in a sentence 112
   in a story arc 96
Results, section of a paper 32, 70
Review paper 32
Reviewer(s)
   as advocate 24, 93, 193
   attitude/patience 19, 31, 40–42,
      70, 183
   expanding the community of 191
Revision 6, 39, 174
Rewriting 6–7
Rules
   breaking 81, 110, 127, 131, 143
   grammatical 112, 192
Rumsfeld, Donald 50

SCFL 175, 189
Schema(s) 20
   and academic disciplines 21, 42
   in describing methods 69
   jargon 147

and learning 20, 23, 39, 105, 113
   nominalizations and 143
Scholar
   developing 39
   vs. expert 195
   sounding scholarly 145
Science (journal) 31, 41, 45
Scientific American 196
Sentence(s) 112–122, 134, 149
   linking sentences 101, 124–129
   long, multiclause 120, 148
   right opening 120
   in story hierarchy 96
   topic sentence 104
Shakespeare, William 158
Shaw, George Bernard 145
"Show, don't tell" 133
Simple ideas, in SUCCES formula 12,
      17–21, 197
Simplification 105, 120
Simplistic 17–18
Society, science and 146, 196
Soil Biology & Biochemistry 18, 116,
      181, 187
Spiral, structure of a story 29, 83, 95
Statistics 76–79
Sticky ideas 16
Story
   arcs 95–103
   elements of 26, 95
   front-loaded 32, 197
   hierarchical structure 96
   modules 24
   in SUCCES formula 24
Storytelling 8, 17, 43, 59, 180, 194
Stream of consciousness writing 102
Stress, of a sentence 113, 118, 126
Strunk, W. Jr. and White, E.B. 22, 104,
      134, 137, 163, 193
Style, writing 24, 44, 146
Subject, of a sentence 56, 112, 134
   long 119, 131, 176
Subject-verb connection 116
SUCCES formula 17
Success, professional 5, 40, 53,
      158, 205
   vs. survival 206

Take-home message 33, 83
Technical terms (vs. jargon) 115, 147, 149–150, 197
Tension 95
Theory, role in science 9, 22, 58, 70
Tolkien, J.R.R 27
Toolbox, writer's 4, 6, 205
Topic sentence 104
Topic, of a sentence 113, 117
Transition(s) 125, 160
TS-D paragraph 104
Twain, Mark 133
Two-Step opening 42

Understanding, as goal of science 10–11, 31–33, 83, 191, 197
Unexpected, in SUCCES formula 21, 24, 30, 50
Unlearning 105, 120

Verb 112
    action 65, 122, 134, 139, 150, 168
    fuzzy 137, 139, 152
    nominalizing 140

Verbosity 167
Vocabulary 146
Voice, writer's 44
    active vs. passive 134
von Moltke, Helmut 180

Watson, J.D. and Crick, F.H.D. 10, 31
Weintraub, Michael 42, 72, 94
Wikipedia 196
Williams, Joseph 7, 105, 113
Word processor 193
Writer
    being a writer 6
    experienced/skilled 14, 102, 155
    inexperienced 39, 46, 119, 193
    professional 3, 133, 205
Writing up, vs. writing 5

Yanai, Ruth 14, 155
"Yes, but" approach to limitations 180

Zinsser, William 7, 29, 30, 35, 47, 124